The publishing house tredition has created the series **TREDITION CLASSICS**. It contains classical literature works from over two thousand years. Most of these titles have been out of print and off the bookstore shelves for decades.

The book series is intended to preserve the cultural legacy and to promote the timeless works of classical literature. As a reader of a **TREDITION CLASSICS** book, the reader supports the mission to save many of the amazing works of world literature from oblivion.

The symbol of **TREDITION CLASSICS** is Johannes Gutenberg (1400 – 1468), the inventor of movable type printing.

With the series, tredition intends to make thousands of international literature classics available in printed format again – worldwide.

All books are available at book retailers worldwide in paperback and in hardcover. For more information please visit: www.tredition.com

tredition was established in 2006 by Sandra Latusseck and Soenke Schulz. Based in Hamburg, Germany, tredition offers publishing solutions to authors and publishing houses, combined with worldwide distribution of printed and digital book content. tredition is uniquely positioned to enable authors and publishing houses to create books on their own terms and without conventional manufacturing risks.

For more information please visit: www.tredition.com

The Chemistry of Hat Manufacturing Lectures Delivered Before the Hat Manufacturers' Association

Watson Smith

Imprint

This book is part of the TREDITION CLASSICS series.

Author: Watson Smith
Cover design: toepferschumann, Berlin (Germany)

Publisher: tredition GmbH, Hamburg (Germany)
ISBN: 978-3-8491-7065-3

www.tredition.com
www.tredition.de

Copyright:
The content of this book is sourced from the public domain.

The intention of the TREDITION CLASSICS series is to make world literature in the public domain available in printed format. Literary enthusiasts and organizations worldwide have scanned and digitally edited the original texts. tredition has subsequently formatted and redesigned the content into a modern reading layout. Therefore, we cannot guarantee the exact reproduction of the original format of a particular historic edition. Please also note that no modifications have been made to the spelling, therefore it may differ from the orthography used today.

PREFACE

The subject-matter in this little book is the substance of a series of Lectures delivered before the Hat Manufacturers' Association in the years 1887 and 1888.

About this period, owing to the increasing difficulties of competition with the products of the German Hat Manufacturers, a deputation of Hat Manufacturers in and around Manchester consulted Sir Henry E. Roscoe, F.R.S., then the Professor of Chemistry in the Owens College, Manchester, and he advised the formation of an Association, and the appointment of a Lecturer, who was to make a practical investigation of the art of Hat Manufacturing, and then to deliver a series of lectures on the applications of science to this industry. Sir Henry Roscoe recommended the writer, then the Lecturer on Chemical Technology in the Owens College, as lecturer, and he was accordingly appointed.

The lectures were delivered with copious experimental illustrations through two sessions, and during the course a patent by one of the younger members became due, which proved to contain the solution of the chief difficulty of the British felt-hat manufacturer (see pages 66-68). This remarkable coincidence served to give especial stress to the wisdom of the counsel [Pg vi] of Sir Henry Roscoe, whose response to the appeal of the members of the deputation of 1887 was at once to point them to scientific light and training as their only resource. In a letter recently received from Sir Henry (1906), he writes: "I agree with you that this is a good instance of the *direct money value* of scientific training, and in these days of 'protection' and similar subterfuges, it is not amiss to emphasise the fact."

It is thus gratifying to the writer to think that the lectures have had some influence on the remarkable progress which the British Hat Industry has made in the twenty years that have elapsed since their delivery.

These lectures were in part printed and published in the *Hatters' Gazette*, and in part in newspapers of Manchester and Stockport, and they have here been compiled and edited, and the necessary illustrations added, etc., by Mr. Albert Shonk, to whom I would express my best thanks.

WATSON SMITH.

London, *April* 1906.

CONTENTS

LECTURE

I. TEXTILE FIBRES, PRINCIPALLY WOOL, FUR, AND HAIR

II. TEXTILE FIBRES, PRINCIPALLY WOOL, FUR, AND HAIR—*continued*

III. WATER: ITS CHEMISTRY AND PROPERTIES; IMPURITIES AND THEIR ACTION; TESTS OF PURITY

IV. WATER: ITS CHEMISTRY AND PROPERTIES; IMPURITIES AND THEIR ACTION; TESTS OF PURITY—*continued*

V. ACIDS AND ALKALIS

VI. BORIC ACID, BORAX, SOAP

VII. SHELLAC, WOOD SPIRIT, AND THE STIFFENING AND PROOFING PROCESS

VIII. MORDANTS: THEIR NATURE AND USE

IX. DYESTUFFS AND COLOURS

X. DYESTUFFS AND COLORS—*continued*

XI. DYEING OF WOOL AND FUR; AND OPTICAL PROPERTIES OF COLOURS

THE CHEMISTRY OF HAT MANUFACTURING

LECTURE I

TEXTILE FIBRES, PRINCIPALLY WOOL, FUR, AND HAIR

Vegetable Fibres.—Textile fibres may be broadly distinguished as vegetable and animal fibres. It is absolutely necessary, in order to obtain a useful knowledge of the peculiarities and properties of animal fibres generally, or even specially, that we should be, at least to some extent, familiar with those of the vegetable fibres. I shall therefore have, in the first place, something to tell you of certain principal vegetable fibres before we commence the more special study of the animal fibres most interesting to you as hat manufacturers, namely, wool, fur, and hair. What cotton is as a vegetable product I shall not in detail describe, but I will refer you to the interesting and complete work of Dr. Bowman, *On the Structure of the Cotton Fibre*. Suffice it to say that in certain plants and trees the seeds or fruit are surrounded, in the pods in which they develop, with a downy substance, and that the cotton shrub belongs to this class of plants. A fibre picked out from the mass of the downy substance referred to, and examined under the microscope, is found to be a spirally twisted band; or better,

[Pg 2]

Fig. 1.

an irregular, more or less flattened and twisted tube (see Fig. 1). We know it is a tube, because on taking a thin, narrow slice across a fibre and examining the slice under the microscope, we can see the hole or perforation up the centre, forming the axis of the tube (see Fig. 2).

Fig. 2.

Mr. H. de Mosenthal, in an extremely interesting and valuable paper (see *J.S.C.I.*, [1] 1904, vol. xxiii. p. 292), has recently shown that the cuticle of the cotton fibre is extremely porous, having, in addition to pores, what appear to be minute stomata, the latter being frequently arranged in oblique rows, as if they led into oblique lateral channels. A cotton fibre varies from 2·5 to 6 centimetres in length, and in breadth from 0·017 to 0·05 millimetre. The characteristics mentioned make it very easy to distinguish cotton from other vegetable or animal fibres. For example, another vegetable fibre is flax, or linen, and this has a very different appearance under the microscope (*see* Fig. 3). It

[Pg 3]

Fig. 3.

has a bamboo-like, or jointed appearance; its tubes are not flattened, nor are they twisted. Flax belongs to a class called the bast fibres, a name given to certain fibres obtained from the inner bark of different plants. Jute also is a bast fibre. The finer qualities of it look like flax, but, as we shall see, it is not chemically identical with cotton, as linen or flax is. Another vegetable fibre, termed "cotton-silk," from its beautiful, lustrous, silky appearance, has excited some attention, because it grows freely in the German colony called the Camaroons, and also on the Gold Coast. This fibre, under the microscope, differs entirely in appearance from both cotton and flax fibres. Its fibres resemble straight and thin, smooth, transparent, almost glassy tubes, with large axial bores; in fact, if wetted in water you can see the water and air bubbles in the tubes under the microscope. A more detailed account of "cotton-silk" appears in a paper read by me before the Society of Chemical Industry in 1886 (see *J.S.C.I.*, 1886, vol. v. p. 642). Now the substance of the cotton, linen or flax, as well as that of the cotton-silk fibres, is termed, chemically, cellulose. Raw cotton consists of cellulose with about 5 per cent. of impurities. This cellulose is a chemical compound of carbon, hydrogen, and oxygen, and, according to the relative proportions of these constituents, it has had the chemical formula $C_6H_{10}O_5$ assigned to it. Each letter [Pg 4] stands for an atom of each constituent named, and the numerals tell us the number of the constituent atoms in the whole compound atom of cellulose. This cellulose is closely allied in composition to starch, dextrin, and a form of sugar called glucose. It is possible to convert cotton rags into this form of sugar—glucose—by treating first with strong vitriol or sulphuric acid, and then boiling with dilute acid for a long time. Before we leave these vegetable or cellulose fibres, I will give you a means of testing them, so as to enable you to distinguish them broadly from the animal fibres, amongst which are silk, wool, fur, and hair. A good general test to distinguish a vegetable and an animal fibre is the following, which is known as Molisch's test: To a very small quantity, about 0·01 gram, of the well-washed cotton fibre, 1 c.c. of water is added, then two to three drops of a 15 to 20 per cent. solution of [Greek: alpha]-naphthol in alcohol, and finally an excess of concentrated sulphuric acid; on agitating, a deep violet colour is developed. By using thymol in place of the [Greek: alpha]-naphthol, a red or scarlet colour is produced. If the fibre were one of an animal nature, merely a yellow or greenish-yellow coloured solution would result. I told you, however, that jute is not chemically identical with cotton and linen. The substance of its fibre has been termed "bastose" by Cross and Bevan, who have investigated it. It is not identical with ordinary cellulose, for if we take a little of the jute, soak it in dilute acid, then in chloride of lime or hypochlorite of soda, and finally pass it through a bath of sulphite of soda, a beautiful crimson colour develops upon it, not developed in the case of cellulose (cotton, linen, etc.). It is certain that it is a kind of cellulose, but still not identical with true cellulose. All animal fibres, when burnt, emit a peculiar empyreumatic odour resembling that from burnt feathers, an odour which no vegetable fibre under like circumstances emits. Hence a good test is to burn a piece of the [Pg 5] fibre in a lamp flame, and notice the odour. All vegetable fibres are easily tendered, or rendered rotten, by the action of even dilute mineral acids; with the additional action of steam, the effect is much more rapid, as also if the fibre is allowed to dry with the acid upon or in it. Animal fibres are not nearly so sensitive under these conditions. But whereas caustic alkalis have not much effect on vegetable fibres, if kept out of contact with the air, the animal fibres are very quickly attacked. Superheated steam alone has but little effect on cotton or vegetable fibres, but it would fuse or melt wool. Based on these differences, methods have been devised and patented for treating mixed woollen and cotton tissues—(1) with hydrochloric acid gas, or moistening with dilute hydrochloric acid and steaming, to remove all the cotton fibre; or (2) with a jet of superheated steam, under a pressure of 5 atmospheres (75 lb. per square inch), when the woollen fibre is simply melted out of the tissue, and sinks to the bottom of the vessel, a vegetable tissue remaining (Heddebault). If we write on paper with dilute sulphuric acid, and dry and then heat the place written upon, the cellulose is destroyed and charred, and we get black writing produced. The principle involved is the same as in the separation of cotton from mixed woollen and cotton goods by means of sulphuric acid or vitriol. The fabric containing cotton, or let us say cellulose particles, is treated with dilute vitriol, pressed or squeezed, and then roughly dried. That cellulose then becomes mere dust, and is simply beaten out of the intact woollen texture. The cellulose is, in a pure state, a white powder, of specific gravity 1·5, *i.e.* one and a half times as heavy as water, and is quite insoluble in such solvents as water, alcohol, ether; but it does dissolve in a solution of hydrated oxide of copper in ammonia. On adding acids to the cupric-ammonium solution, the cellulose is reprecipitated in the form of a gelatinous mass. Cotton and linen are scarcely dissolved at all by a solution of basic zinc chloride.

[1] *J.S.C.I.* = *Journal of the Society of Chemical Industry.*

[Pg 6]

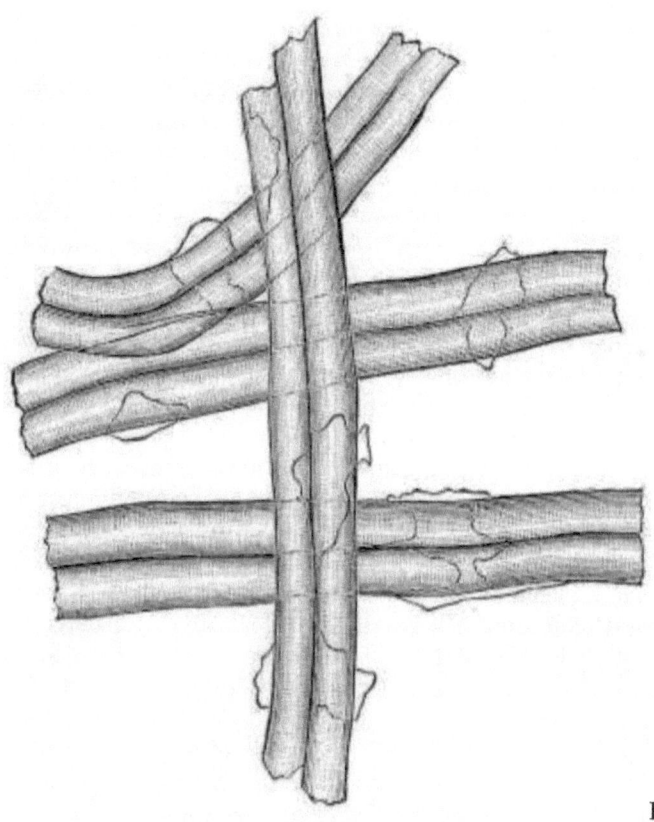

Fig. 4.

Silk. — We now pass on to the animal fibres, and of these we must first consider silk. This is one of the most perfect substances for use in the textile arts. A silk fibre may be considered as a kind of rod of solidified flexible gum, secreted in and exuded from glands placed on the side of the body of the silk-worm. In Fig. 4 are shown the forms of the silk fibre, in which there are no central cavities or axial bores as in cotton and flax, and no signs of any cellular structure or external markings, but a comparatively smooth, glassy surface. There is, however, a longitudinal groove of more or less depth. The fibre is semi-transparent, the beautiful pearly lustre being due to the smoothness of the outer layer and its reflection of the light. In the silk fibre there are two distinct parts: first, the central portion, or, as we may regard it, the true fibre, [Pg 7] chemically termed *fibroïn*; and secondly, an envelope composed of a substance or substances, chemically termed *sericin*, and often "silk-glue" or "silk-gum." Both the latter and *fibroïn* are composed of carbon, hydrogen, nitrogen, and oxygen. Here there is thus one element more than in the vegetable fibres previously referred to, namely, nitrogen; and this nitrogen is contained in all the animal fibres. The outer envelope of silk-glue or sericin can be dissolved off the inner fibroïn fibre by means of hot water, or warm water with a little soap. Warm dilute (that is, weak) acids, such as sulphuric acid, etc., also dissolve this silk-glue, and can be used like soap solutions for ungumming silk. Dilute nitric acid only slightly attacks silk, and colours it yellow; it would not so colour vegetable fibres, and this forms a good test to distinguish silk from a vegetable fibre. Cold strong acetic acid, so-called glacial acetic acid, removes the yellowish colouring matter from raw silk without dissolving the sericin or silk-gum. By heating under pressure with acetic acid, however, silk is completely dissolved. Silk is also dissolved by strong sulphuric acid, forming a brown thick liquid. If we add water to this thick liquid, a clear solution is obtained, and then on adding tannic acid the fibroïn is precipitated. Strong caustic potash or soda dissolves silk; more easily if warm. Dilute caustic alkalis, if sufficiently dilute, will dissolve off the sericin and leave the inner fibre of fibroïn; but they are not so good for ungumming silk as soap solutions are, as the fibre after treatment with them is deficient in whiteness and brilliancy. Silk dissolves completely in hot basic zinc chloride solution, and also in an alkaline solution of copper and glycerin, which solutions do not dissolve vegetable fibres or wool. Chlorine and bleaching-powder solutions soon attack and destroy silk, and so another and milder agent, namely, sulphurous acid, is used to bleach this fibre. Silk is easily dyed by the aniline and coal-tar colours, and with beautiful effect, but it has little attraction for the mineral colours.

[Pg 8]

Wool.—Next to silk as an animal fibre we come to wool and different varieties of fur and hair covering certain classes of animals, such as sheep, goats, rabbits, and hares. Generally, and without going at all deeply into the subject, we may say that wool differs from fur and hair, of which we may regard it as a variety, by being usually more elastic, flexible, and curly, and because it possesses certain features of surface structure which confer upon it the property of being more easily matted together than fur and hair are. We must first shortly consider the manner of growth of hair without spending too much time on this part of the subject. The accompanying figure (see Fig. 5) shows a section of the skin with a hair or wool fibre rooted in it. Here we may see that the ground work, if we may so term it, is four-fold in structure. Proceeding downwards, we have—(first) the outer skin, scarf-skin or cuticle; (second) a second layer or skin called the *rete mucosum*, forming the epidermis; (third) papillary layer; (fourth) the corium layer, forming the dermis. The peculiar, globular, cellular masses below in the corium are called adipose cells, and these throw off perspiration or moisture, which is carried away to the surface by the glands shown (called sudoriparous glands), which, as is seen, pass independently off to the surface. Other glands terminate under the skin in the hair follicles, which follicles or hair sockets contain or enclose the hair roots. These glands terminating in the hair follicles secrete an oily substance, which bathes and lubricates as well as nourishes the hair. With respect to the origin of the hair or wool fibre, this is formed inside the follicle by the exuding therefrom of a plastic liquid or lymph; this latter gradually becomes granular, and is then formed into cells, which, as the growth proceeds, are elongated into fibres, which form the central portion of the hair. Just as with the trunk of a tree, we have an outer dense portion, the bark, an inner less dense and more cellular layer, and an inmost

[Pg 9]

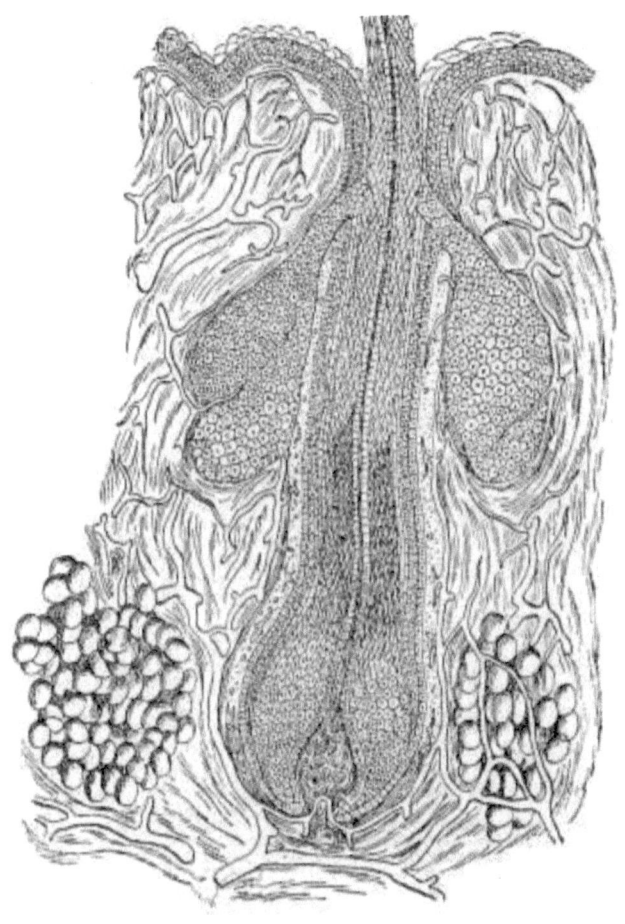

Fig. 5.

portion which is most cellular and porous; so with a hair, the central portion is loose and porous, the outer more and more dense. On glancing at the figure (Fig. 6) of the longitudinal section of a human hair, we see first the outer portion, like the bark of a tree, consisting of a dense sheath of flattened scales, then comes an inner lining of closely-packed fibrous

[Pg 10]

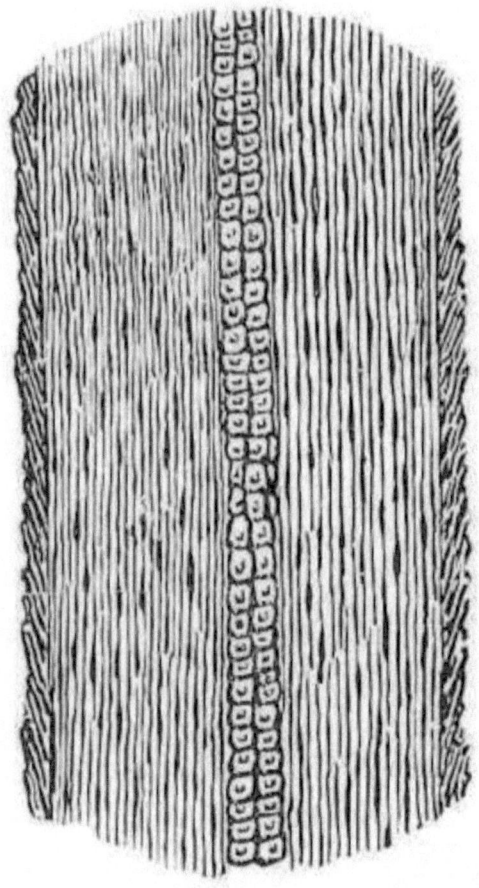

Fig. 6.

cells, and frequently an inner well-marked central bundle of larger and rounder cells, forming a medullary axis. The transverse section (Fig. 7) shows this exceedingly well. The end of a hair is generally pointed, sometimes filamentous. The lower extremity is larger than the shaft, and terminates in a conical bulb, or mass of cells, which forms the root of the

[Pg 11]

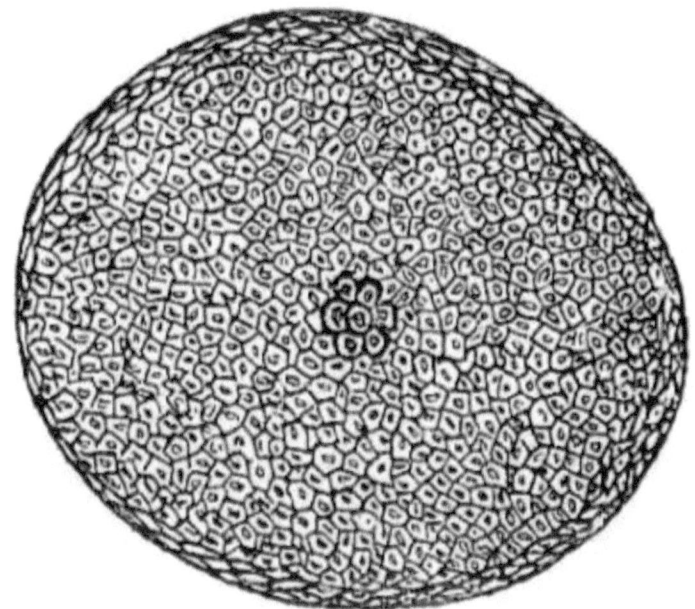

Fig. 7.

hair. In the next figure (Fig. 8) we are supposed to have separated these cells, and above, (a), we see some of the cells from the central pith or medulla, and fat globules; between, (b), some of the intermediate elongated or angular cells; and below, (c), two flattened, compressed, structureless, and horny scales from the outer portion of the hair. Now these latter flattened scales are of great importance.

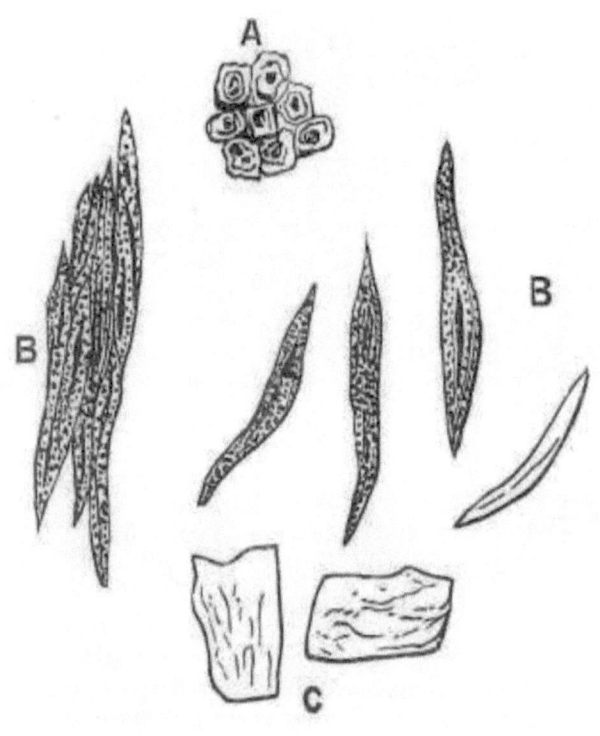

Fig. 8.

Their character and mode of connection with the stratum, or cortical substance, below, not only make all the difference between wool and hair, but also determine the extent and degree of that peculiar property of interlocking of the hairs known as felting. Let us now again look at a human hair. The light was reflected from this hair as it lay under the microscope, and now we see the reason of the saw-like edge in the longitudinal section, for just as the tiles lie on the roof of a house, or the scales on the back of a [Pg 12] fish, so the whole surface of the hair is externally coated with a firmly adhering layer of flat overlying scales, with not very even upper edges, as you see. The upper or free edges of these scales are all directed towards the end of the hair, and away from the root. But when you look at a hair in its natural state you cannot see these scales, so flat do they lie on the hair-shaft. What you see are only irregular transverse lines across it. Now I come to a matter of great importance, as will later on appear in connection with means for promoting felting properties. If a hair such as described, with the scales lying flat on the shaft, be treated with certain substances or reagents which act upon and dissolve, or decompose or disintegrate its parts, then the free edges of these scales rise up, they "set their backs up," so to say. They, in fact, stand off like the scales of a fir-cone, and at length act like the fir-cone in ripening, at last becoming entirely loose. As regards wool and fur, these scales are of the utmost importance, for very marked differences exist even in the wool of a single sheep, or the fur of a single hare. It is the duty of the wool-sorter to distinguish and separate the various qualities in each fleece, and of the furrier to do the same in the case of each fur. In short, upon the nature and arrangement and conformation of the scales on the hair-shafts, especially as regards those free upper edges, depends the distinction of the value of many classes of wool and fur. These scales vary both as to nature and arrangement in the case of the hairs of different animals, so that by the aid of the microscope we have often a means of determining from what kind of animal the hair has been derived. It is on the nature of this outside scaly covering of the shaft, and in the manner of attachment of these scaly plates, that the true distinction between wool and hair rests. The principal epidermal characteristic of a true wool is the capacity of its fibres to felt or mat together. This arises from the greater looseness of

[Pg 13]

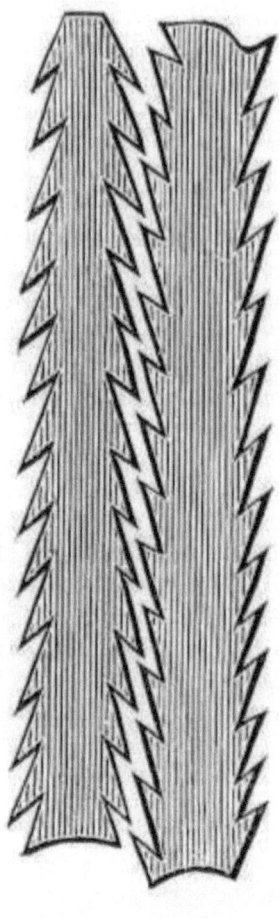

Fig. 9.

the scaly covering of the hair, so that when opposing hairs come into contact, the scales interlock (see Fig. 9), and thus the fibres are held together. Just as with hair, the scales of which have their free edges pointing upwards away from the root, and towards the extremity of the hair, so with wool. When the wool is on the back of the sheep, the scales of the woolly hair all point in the same direction, so that while maintained in that attitude the individual hairs slide over one another, and do not tend to felt or mat; if they did, woe betide the animal. The fact of the peculiar serrated, scaly structure of hair and wool is easily proved by working a hair between the fingers. If, for instance, a human hair be placed between finger and thumb, and gently rubbed by the alternate motion of finger and thumb together, it will then invariably move in the direction of the root, quite independently of the will of the person performing the test. A glance at the form of the typical wool fibres shown (see Fig. 10),

[Pg 14]

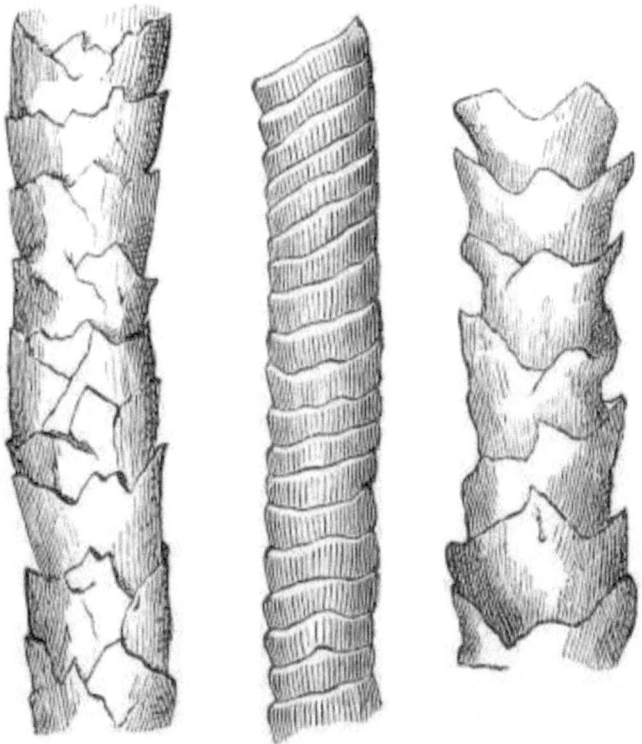

Fig. 10.
Finest merino wool fibre. Typical wool fibre. Fibre of wool from Chinese sheep.

will show the considerable difference between a wool and a hair fibre. You will observe that the scales of the wool fibre are rather pointed than rounded at their free edges, and that at intervals we have a kind of composite and jagged-edged funnels, fitting into each other, and thus making up the covering of the cylindrical portion of the fibre. The sharpened, jagged edges enable these scales more easily to get under the opposing scales, and to penetrate inwards and downwards according to the pressure exerted. The free edges of the scales of wool are much longer and deeper than in the case of hair. In hair the overlapping scales are attached to the under layer up to the edges of those scales, and at this extremity can only be detached [Pg 15] by the use of certain reagents. But this is not so with wool, for here the ends of the scales are, for nearly two-thirds of their length, free, and are, moreover, partially turned outwards. One of the fibres shown in Fig. 10 is that of the merino sheep, and is one of the most valuable and beautiful wools grown. There you have the type of a fibre best suited for textile purposes, and the more closely different hairs approach this, the more suitable and valuable they become for those purposes, and *vice versâ*. With regard to the curly structure of wool, which increases the matting tendency, though the true cause of this curl is not known, there appears to be a close relationship between the tendency to curl, the fineness of the fibre, and the number of scales per linear inch upon the surface. With regard to hair and fur, I have already shown that serrated fibres are not specially peculiar to sheep, but are much more widely diffused. Most of the higher members of the mammalia family possess a hairy covering of some sort, and in by far the larger number is found a tendency to produce an undergrowth of fine woolly fibre, especially in the winter time. The differences of human hair and hairs generally, from the higher to the lower forms of mammalia, consist only in variations of size and arrangement as regards the cells composing the different parts of the fibre, as well as in a greater or less development of the scales on the covering or external hair surface. Thus, under the microscope, the wool and hairs of various animals, as also even hairs from different parts of the same animal, show a great variety of structure, development, and appearance.

We have already observed that hair, if needed for felting, is all the better—provided, of course, no injury is done to the fibre itself—for some treatment, by which the scales otherwise lying flatter on the hair-shafts than in the case of the hairs of wool, are made to stand up somewhat, extending outwards their free edges. This brings me to the consideration of a practice pursued by [Pg 16] furriers for this purpose, and known as the *sécretage* or "carrotting" process; it consists in a treatment with a solution of mercuric nitrate in nitric acid, in order to improve the felting qualities of the fur. This acid mixture is brushed on to the fur, which is cut from the skin by a suitable sharp cutting or shearing machine. A Manchester furrier, who gave me specimens of some fur untreated by the process, and also some of the same fur that had been treated, informed me that others of his line of business use more mercury than he does, *i.e.* leave less free nitric acid in their mixture; but he prefers his own method, and thinks it answers best for the promotion of felting. The treated fur he gave me was turned yellow with the nitric acid, in parts brown, and here and there the hairs were slightly matted with the acid. In my opinion the fur must suffer from such unequal treatment with such strong acid, and in the final process of finishing I should not be surprised if difficulty were found in getting a high degree of lustre and finish upon hairs thus roughened or partially disintegrated. Figs. 11 and 12 respectively illustrate fur fibres from different parts of the same hare before and after the treatment. In examining one of these fibres from the side of a hare, you see what the cause of this roughness is, and what is also the cause of the difficulty in giving a polish or finish. The free edges are partially disintegrated, etched as it were, besides being caused to stand out. A weaker acid ought to be used, or more mercury and less acid. As we shall afterwards see, another dangerous agent, if not carefully used, is bichrome (bichromate of potassium), which is also liable to roughen and injure the fibre, and thus interfere with the final production of a good finish.

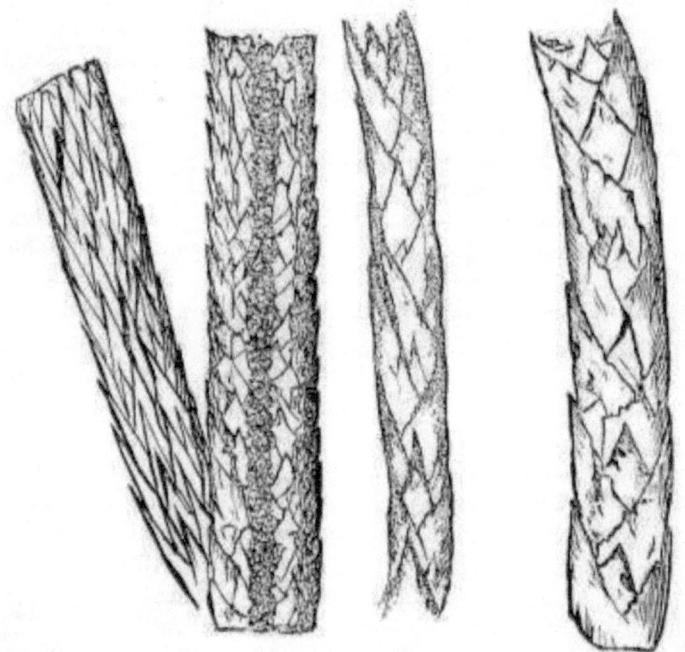

Fig. 11.

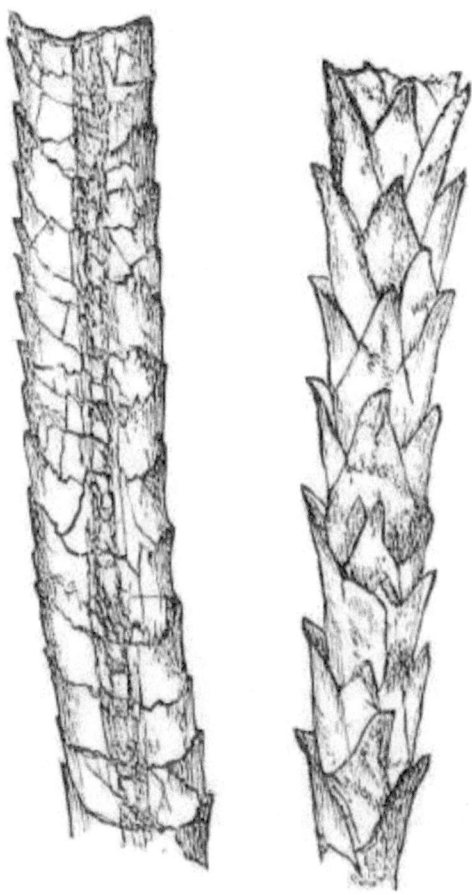

Fig. 12.

[Pg 17]

LECTURE II

TEXTILE FIBRES, PRINCIPALLY WOOL, FUR, AND HAIR — *Continued*

With regard to the preparation of fur by acid mixtures for felting, mentioned in the last lecture, I will tell you what I think I should recommend. In all wool and fur there is a certain amount of grease, and this may vary in different parts of the material. Where there is most, however, the acid, nitric acid, or nitric acid solution of nitrate of mercury, will wet, and so act on the fur, least. But the action ought to be uniform, and I feel sure it cannot be until the grease is removed. I should therefore first wash the felts on the fur side with a weak alkaline solution, one of carbonate of soda, free from any caustic, to remove all grease, then with water to remove alkali; and my belief is that a weaker and less acid solution of nitric acid and nitrate of mercury, and a smaller quantity of it, would then do the work required, and do it more uniformly.

A question frequently asked is: "Why will dead wool not felt?" Answer: If the animal become weak and diseased, the wool suffers degradation; also, with improvement in health follows *pari passu*, improvement in the wool structure, which means increase both in number and vigour of the scales on the wool fibres, increase of the serrated ends of these, and of their regularity. In weakness and disease the number of scales in a given hair-shaft diminishes, and these become finer and less

[Pg 18]

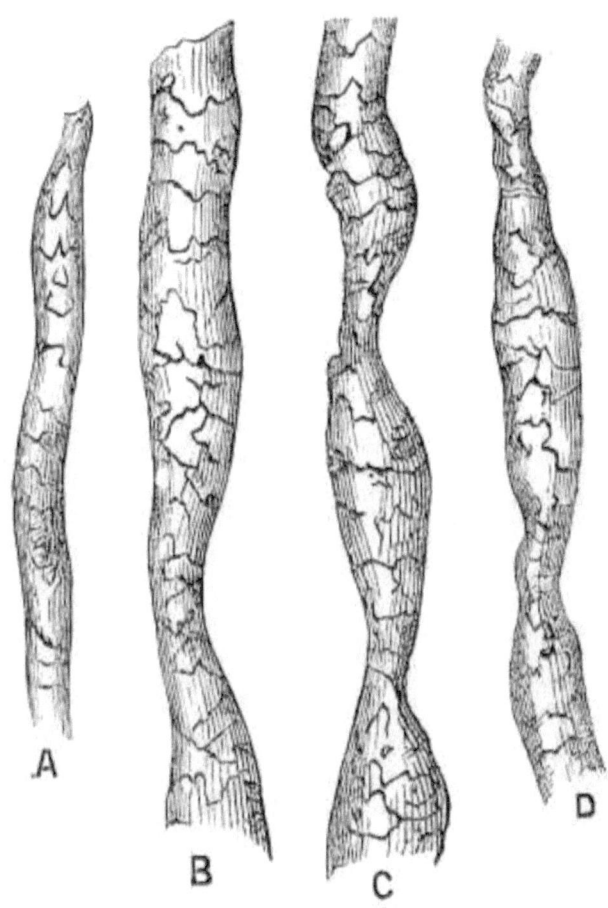

Fig. 13.

pronounced. The fibres themselves also become attenuated. Hence when disease becomes death, we have considerably degraded fibres. This is seen clearly in the subjoined figures (see Fig. 13), which are of wool fibres from animals that have died of disease. The fibres are attenuated and irregular, the scale markings and edges have almost disappeared in some places, and are generally scanty and meagre in development. It is no wonder that such "dead wool" will be badly adapted for felting. "Dead wool" is nearly as bad as "kempy" wool, in which malformation of fibre has occurred. In such "kemps," as Dr. Bowman has shown, scales have disappeared, and the fibre has become, in part or whole, a dense, non-cellular structure, resisting dye-penetration and felting (see Fig. 14).

[Pg 19]

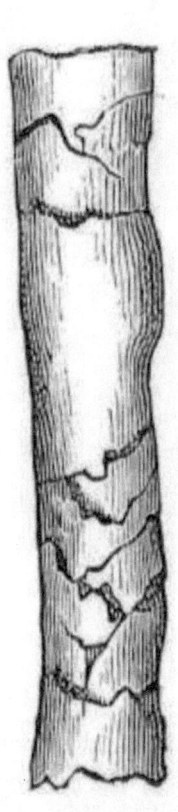

Fig. 14.

One of the physical properties of wool is its hygroscopicity or power of absorbing moisture. As the very structure of wool and fur fibre would lead us to suppose, these substances are able to absorb a very considerable amount of water without appearing damp. If exposed freely to the air in warm and dry weather, wool retains from 8 to 10 per cent., and if in a damp place for some time, it may absorb as much as from 30 to 50 per cent. of water: Wool, fur, or hair that has been washed, absorbs the most moisture; indeed, the amount of water taken up varies inversely with the fatty or oily matter present. Hence the less fat the more moisture. In the washed wool, those fibres in which the cells are more loosely arranged have the greatest absorbing power for water. No doubt the moisture finds its way in between the cells of the wool fibre from which the oil or fat has been removed. But I need hardly remind you that if wool and fur are capable, according to the circumstances under which they are placed, of absorbing so much moisture as that indicated, [Pg 20] it becomes (especially in times of pressure and competition) very important to inquire if it be not worth while to cease paying wool and fur prices for mere water. This question was answered long ago in the negative by our Continental neighbours, and in Germany, France, and Switzerland official conditioning establishments have been founded by the Governments of those countries for the purpose of testing lots of purchased wool and silk, etc., for moisture, in order that this moisture may be deducted from the invoices, and cash paid for real dry wool, etc. I would point out that if you, as hat manufacturers, desire to enter the lists with Germany, you must not let her have any advantage you have not, and it is an advantage to pay for what you know exactly the composition of, rather than for an article that may contain 7 per cent. or, for aught you know, 17 per cent. or 30 per cent. of water. There is, so far as I know, no testing for water in wools and furs in this country, and certainly no "conditioning establishments" (1887), and, I suppose, if a German or French wool merchant or furrier could be imagined as selling wool, etc., in part to a German or French firm, and in part to an English one, the latter would take the material without a murmur, though it might contain 10 per cent., or, peradventure, 30 per cent. of water, and no doubt the foreign, just as the English merchant or dealer, would get the best price he could, and regard the possible 10 per cent. or 30 per cent. of water present with certainly the more equanimity the more of that very cheap element there were present. But look at the other side. The German or French firm samples its lot as delivered, takes the sample to be tested, and that 10 or 30 per cent. of water is deducted, and only the dry wool is paid for. A few little mistakes of this kind, I need hardly say, will altogether form a kind of *vade mecum* for the foreign competitor.

We will now see what the effect of water is in the felting operation. Especially hot water assists that operation, and [Pg 21] the effect is a curious one. When acid is added as well, the felting is still further increased, and shrinking also takes place. As already shown you, the free ends of the scales, themselves softened by the warm dilute acid, are extended and project more, and stand out from the shafts of the hairs. On the whole, were I a hat manufacturer, I should prefer to buy my fur untreated by that nitric acid and mercury process previously referred to, and promote its felting properties myself by the less severe and more rational course of proceeding, such, for example, as treatment with warm dilute acid. We have referred to two enemies standing in the way to the obtainment of a final lustre and finish on felted wool or fur, now let us expose a third. In the black dyeing of the hat-forms a boiling process is used. Let us hear what Dr. Bowman, in his work on the wool fibre, says with regard to boiling with water. "Wool which looked quite bright when well washed with tepid water, was decidedly duller when kept for some time in water at a temperature of 160° F., and the same wool, when subjected to boiling water at 212° F., became quite dull and lustreless. When tested for strength, the same fibres which carried on the average 500 grains without breaking before boiling, after boiling would not bear more than 480 grains." Hence this third enemy is a boiling process, especially a long-continued one if only with water itself. If we could use coal-tar colours and dye in only a warm weak acid bath, not boil, we could get better lustre and finish.

We will now turn our attention to the chemical composition of wool and fur fibres. On chemical analysis still another element is found over and above those mentioned as the constituents of silk fibre. In silk, you will recollect, we observed the presence of carbon, hydrogen, oxygen, and nitrogen. In wool, fur, etc., we must add a fifth constituent, namely, sulphur. Here is an analysis of pure German wool—Carbon, [Pg 22] 49·25 per cent.; hydrogen, 7·57; oxygen, 23·66; nitrogen, 15·86; sulphur, 3·66—total, 100·00. If you heat either wool, fur, or hair to 130° C., it begins to decompose, and to give off ammonia; if still further heated to from 140° to 150° C., vapours containing sulphur are evolved. If some wool be placed in a dry glass tube, and heated strongly so as to cause destructive distillation, products containing much carbonate of ammonium are given off. The ammonia is easily detected by its smell of hartshorn and the blue colour produced on a piece of reddened litmus paper, the latter being a general test to distinguish alkalis, like ammonia, soda, and potash, from acids. No vegetable fibres will, under any circumstances, give off ammonia. It may be asked, "But what does the production of ammonia prove?" I reply, the "backbone," chemically speaking, of ammonia is nitrogen. Ammonia is a compound of nitrogen and hydrogen, and is formulated NH_3, and hence to discover ammonia in the products as mentioned is to prove the prior existence of its nitrogen in the wool, fur, and hair fibres.

Action of Acids on Wool, etc.—Dilute solutions of vitriol (sulphuric acid) or hydrochloric acid (muriatic acid, spirits of salt) have little effect on wool, whether warm or cold, except to open out the scales and confer roughness on the fibre. Used in the concentrated state, however, the wool or fur would soon be disintegrated and ruined. But under all circumstances the action is far less than on cotton, which is destroyed at once and completely. Nitric acid acts like sulphuric and hydrochloric acids, but it gives a yellow colour to the fibre. You see this clearly enough in the fur that comes from your furriers after the treatment they subject it to with nitric acid and nitrate of mercury. There is a process known called the stripping of wool, and it consists in destroying the colour of wool and woollen goods already dyed, in order that they may be re-dyed. Listen, however, to the important precautions followed: A nitric acid not stronger [Pg 23] than from 3° to 4° Twaddell is used, and care is taken not to prolong the action more than three or four minutes.

Action of Alkalis.—Alkalis have a very considerable action on fur and wool, but the effects vary a good deal according to the kind of alkali used, the strength and the temperature of the solution, as also, of course, the length of period of contact. The caustic alkalis, potash and soda, under all conditions affect wool and fur injuriously. In fact, we have a method of recovering indigo from indigo-dyed woollen rags, based on the solubility of the wool in hot caustic soda. The wool dissolves, and the indigo, being insoluble, remains, and can be recovered. Alkaline carbonates and soap in solution have little or no injurious action if not too strong, and if the temperature be not over 50° C. (106° F.). Soap and carbonate of ammonium have the least injurious action. Every washer or scourer of wool, when he uses soaps, should first ascertain if they are free from excess of alkali, *i.e.* that they contain no free alkali; and when he uses soda ash (sodium carbonate), that it contains no caustic alkali. Lime, in water or otherwise, acts injuriously, rendering the fibre brittle.

Reactions and tests proving chemical differences and illustrating modes of discriminating and separating vegetable fibres, silk and wool, fur, etc. — You will remember I stated that the vegetable fibre differs chemically from those of silk, and silk from wool, fur, and hair, in that with the first we have as constituents only carbon, hydrogen, and oxygen; in silk we have carbon, hydrogen, oxygen, and nitrogen; whilst in wool, fur, and hair we have carbon, hydrogen, oxygen, nitrogen, and sulphur. I have already shown you that if we can liberate by any means ammonia from a substance, we have practically proved the presence of nitrogen in that substance, for ammonia is a nitrogen compound. As regards sulphur and its compounds, that ill-smelling gas, sulphuretted hydrogen, which occurs in rotten eggs, in organic effluvia from cesspools and the [Pg 24] like, and which in the case of bad eggs, and to some extent with good eggs, turns the silver spoons black, and in the case of white lead paints turns these brown or black, I can show you some still more convincing proofs that sulphur is contained in wool, fur, and hair, and not in silk nor in vegetable fibres. First, I will heat strongly some cotton with a little soda-lime in a tube, and hold a piece of moistened red litmus paper over the mouth of the tube. If nitrogen is present it will take up hydrogen in the decomposition ensuing, and escape as ammonia, which will turn the red litmus paper blue. With the cotton, however, no ammonia escapes, no turning of the piece of red litmus paper blue is observed, and so no nitrogen can be present in the cotton fibre. Secondly, I will similarly treat some silk. Ammonia escapes, turns the red litmus paper blue, possesses the smell like hartshorn, and produces, with hydrochloric acid on the stopper of a bottle, dense white fumes of sal-ammoniac (ammonium chloride). Hence silk contains nitrogen. Thirdly, I will heat some fur with soda-lime. Ammonia escapes, giving all the reactions described under silk. Hence fur, wool, etc., contain nitrogen. As regards proofs of all three of these classes of fibres containing carbon, hydrogen, and oxygen, the char they all leave behind on heating in a closed vessel is the carbon itself present. For the hydrogen and oxygen, a perfectly dry sample of any of these fabrics is taken, of course in quantity, and heated strongly in a closed vessel furnished with a condensing worm like a still. You will find all give you water as a condensate—the vegetable fibre, acid water; the animal fibres, alkaline water from the ammonia. The presence of water proves both hydrogen and oxygen, since water is a compound of these elements. If you put a piece of potassium in contact with the water, the latter will at once decompose, the potassium absorbing the oxygen, and setting free the hydrogen as gas, which you could collect and ignite with a match, when you would find it would [Pg 25] burn. That hydrogen was the hydrogen forming part of your cotton, silk, or wool, as the case might be. We must now attack the question of sulphur. First, we prepare a little alkaline lead solution (sodium plumbate) by adding caustic soda to a solution of lead acetate or sugar of lead, until the white precipitate first formed is just dissolved. That is one of our reagents; the other is a solution of a red-coloured salt called nitroprusside of sodium, made by the action of nitric acid on sodium ferrocyanide (yellow prussiate). The first-named is very sensitive to sulphur, and turns black directly. To show this, we take a quantity of flowers of sulphur, dissolve in caustic soda, and add to the lead solution. It turns black at once, because the sulphur unites with the lead to form black sulphide of lead. The nitroprusside, however, gives a beautiful crimson-purple coloration. Now on taking a little cotton and heating with the caustic alkaline lead solution, if sulphur were present in that cotton, the fibre would turn black or brown, for the lead would at once absorb such sulphur, and form in the fibre soaked with it, black sulphide of lead. No such coloration is formed, so cotton does not contain sulphur. Secondly, we must test silk. Silk contains nitrogen, like wool, but does it contain sulphur? The answer furnished by our tests is— no! since the fibre is not coloured brown or black on heating with the alkaline lead solution. Thirdly, we try some white Berlin wool, so that we can easily see the change of colour if it takes place. In the hot lead solution the wool turns black, lead sulphide being formed. On adding the nitroprusside solution to a fresh portion of wool boiled with caustic soda, to dissolve out the sulphur, a splendid purple coloration is produced. Fur and hair would, of course, do the same thing. Lead solutions have been used for dyeing the hair black; not caustic alkaline solutions like this, however. They would do something more than turn the hair black—probably give rise to some vigorous exercise of muscular power! [Pg 26] Still it has been found that even the lead solutions employed have, through gradual absorption into the system, whilst dyeing the hair black, also caused colics and contractions of the limbs.

Having now found means for proving the presence of the various elements composing cotton, silk, and wool, fur or hair, we come to methods that have been proposed for distinguishing these fibres more generally, and for quantitatively determining them in mixtures. One of the best of the reagents for this purpose is the basic zinc chloride already referred to. This is made as follows: 100 parts of fused zinc chloride, 85 parts of water, and 4 parts of zinc oxide are boiled together until a clear solution is obtained. This solution dissolves silk slowly in the cold, quickly if hot, and forms a thick gummy liquid. Wool, fur, and vegetable fibres are not affected by it. Hence if we had a mixture, and treated with this solution, we could strain off the liquid containing the dissolved silk, and would get cotton and wool left. On weighing before and after such treatment, the difference in weights would give us the silk present. The residue boiled with caustic soda would lose all its wool, which is soluble in hot strong caustic alkali. Again straining off, we should get only the cotton or other vegetable fibre left, and thus our problem would be solved. Of course there are certain additional niceties and modifications still needed, and I must refer you for the method in full to the *Journal of the Society of Chemical Industry*, 1882, page 64; also 1884, page 517. I will now conclude with some tests with alkaline and acid reagents, taken in order, and first the acids. These will also impress upon our minds the effects of acids and alkalis on the different kinds of fibres.

I. In three flasks three similar portions of cotton lamp-wick, woollen yarn, and silk are placed, after previously moistening them in water and wringing them out. To each is now added similar quantities of concentrated sulphuric acid. The cotton is quickly broken up and dissolved, especially if assisted by [Pg 27] gentle warming, and at last a brown, probably a black-brown, solution is obtained. The woollen is a little broken up, but not much to the naked eye, and the vitriol is not coloured. The silk is at once dissolved, even in the cold acid. We now add excess of water to the contents of each flask. A brownish, though clear, solution is produced in the case of cotton; the woollen floats not much injured in the acid, whilst a clear limpid solution is obtained with the silk. On adding tannic acid solution to all three, only the silk yields a precipitate, a rather curdy one consisting of fibroïn.

II. Three specimens of cotton, wool, and silk, respectively, are touched with nitric acid. Cotton is not coloured, but wool and silk are stained yellow; they are practically dyed.

III. Three specimens, of cotton, wool, and silk, respectively, are placed in three flasks, and caustic soda solution of specific gravity 1·05 (10° Twaddell) is added. On boiling, the wool and silk dissolve, whilst the cellulose fibre, cotton, remains undestroyed.

IV. If, instead of caustic soda as in III., a solution of oxide of copper in ammonia be used, cotton and silk are dissolved, but wool remains unchanged, *i.e.* undissolved. If sugar or gum solutions be added to the solutions of cotton and silk, the cotton cellulose is precipitated, whilst the silk is not, but remains in solution.

V. Another alkaline solvent for silk, which, however, leaves undissolved cotton and wool, is prepared as follows: 16 grains of copper sulphate ("blue vitriol," "bluestone") are dissolved in 150 c.c. of water, and then 16 grains of glycerin are added. To this mixture a solution of caustic soda is added until the precipitate first formed is just re-dissolved, so as not to leave an excess of caustic soda present.

[Pg 28]

LECTURE III

WATER: ITS CHEMISTRY AND PROPERTIES; IMPURITIES AND THEIR ACTION; TESTS OF PURITY

I have already had occasion to refer, in my last Lecture, to water as a chemical substance, as a compound containing and consisting of hydrogen and oxygen. What are these water constituents, hydrogen and oxygen? Each of them is a gas, but each a gas having totally different properties. On decomposing water and collecting the one of these two gases, the hydrogen gas, in one vessel, and the other, the oxygen gas, in another vessel, twice as large a volume of hydrogen gas is given off by the decomposing water as of oxygen. You may now notice a certain meaning in the formula assigned to water, H_2O: two volumes of hydrogen combined with one of oxygen; and it may be added that when such combination takes place, not three volumes of resulting water vapour (steam), but two volumes are produced. This combination of the two gases, when mixed together, is determined by heating to a high temperature, or by passing an electric spark; it then takes place with the consequent sudden condensation of three volumes of mixture to two of compound, so as to cause an explosion. I may also mention that as regards the weights of these bodies, oxygen and hydrogen, the first is sixteen times as heavy as the second; and since we adopt hydrogen as the unit, we may consider H to stand for hydrogen, and also to signify 1 — the unit; whilst O means [Pg 29] oxygen, and also 16. Hence the compound atom or molecule of water, H_2O, weighs 18. I must now show you that these two gases are possessed of totally different properties. Some gases will extinguish a flame; some will cause the flame to burn brilliantly, but will not burn themselves; and some will take fire and burn themselves, though extinguishing the flame which has ignited them. We say the first are non-combustible, and will not support combustion; the second are supporters of combustion, the third are combustible gases. Of course these are, as the lawyers say, only *ex parte* statements of the truth; still they are usually accepted. Oxygen gas will ignite a red-hot match, but hydrogen will extinguish an inflamed one, though it will itself burn. You generally think of water as the great antithesis of, the universal antidote for, fire. The truth is here again only of an *ex parte* character, as I will show you. If I can, by means of a substance having a more intense affinity for oxygen than hydrogen has, rob water of its oxygen, I necessarily set the hydrogen that was combined with that oxygen free. If the heat caused by the chemical struggle, so to say, is great, that hydrogen will be inflamed and burn. Thus we are destroying that antithesis, we are causing the water to yield us fire. I will do this by putting potassium on water, and even in the cold this potassium will seize upon the oxygen of the water, and the hydrogen will take fire.

Specific Gravity. — We must now hasten to other considerations of importance. Water is generally taken as the unit in specific gravities assigned to liquids and solids. This simply means that when we desire to express how heavy a thing is, we are compelled to say it is so many times heavier or lighter than something. That something is generally water, which is regarded, consequently, as unit or figure 1. A body of specific gravity 1·5, or 1½, means that that body is 1½ or 1·5 times as heavy as water. As hat manufacturers, you will have mostly to do with the specific gravities of liquids, aqueous solutions, [Pg 30] and you will hear more of Twaddell degrees. The Twaddell hydrometer, or instrument for measuring the specific gravities of liquids, is so constructed that when it stands in water, the water is just level with its zero or 0° mark. Well, since in your reading of methods and new processes, you will often meet with specific gravity numbers and desire to convert these into Twaddell degrees, I will give you a simple means of doing this. Add cyphers so as to make into a number of four figures, then strike out the unit and decimal point farthest to the left, and divide the residue by 5, and you get the corresponding Twaddell degrees. If you have Twaddell degrees, simply multiply by 5, and add 1000 to the result, and you get the specific gravity as usually taken, with water as the unit, or in this case as 1000. An instrument much used on the Continent is the Beaumé hydrometer. The degrees (n) indicated by this instrument can be converted into specific gravity (d) by the formula: $d = 144\cdot 3/(144\cdot 3 - n)$

Ebullition or Boiling of Water, Steam. — The atmosphere around us is composed of a mixture of nitrogen and oxygen gases; not a compound of these gases, as water is of hydrogen and oxygen, but a mixture more like sand and water or smoke and air. This mass of gases has weight, and presses upon objects at the surface of the earth to the extent of 15 lb. on the square inch. Now some liquids, such as water, were it not for this atmospheric pressure, would not remain liquids at all, but would become gases. The pressure thus tends to squeeze gases together and convert them into liquids. Any force that causes gases to contract will do the same thing, of course—for example, cold; and *ceteris paribus* removal of pressure and expansion by heat will act so as to gasify liquids. When in the expansion of liquids a certain stage or degree is reached, different for different liquids, gas begins to escape so quickly from the liquid that bubbles of vapour are continually formed [Pg 31] and escape. This is called ebullition or boiling. A certain removal of pressure, or expansion by heat, is necessary to produce this, *i.e.* to reach the boiling-point of the liquid. As regards the heat necessary for the boiling of water at the surface of the earth, *i.e.* under the atmospheric pressure of 15 lb. on the square inch, this is shown on the thermometer of Fahrenheit as 212°, and on the simpler centigrade one, as 100°, water freezing at 0° C. But if what I have said is true, when we remove some of the atmospheric pressure, the water should boil with a less heat than will cause the mercury in the thermometer to rise to 100° C., and if we take off all the pressure, the water ought to boil and freeze at the same time. This actually happens in the Carré ice-making machine. The question now arises, "Why does the water freeze in the Carré machine?" All substances require certain amounts of heat to enable them to take and to maintain the liquid state if they are ordinarily solid, and the gaseous state if ordinarily liquid or solid, and the greater the change of state the greater the heat needed. Moreover, this heat does not make them warm, it is simply absorbed or swallowed up, and becomes latent, and is merely necessary to maintain the new condition assumed. In the case of the Carré machine, liquid water is, by removal of the atmospheric pressure, coerced, as it were, to take the gaseous form. But to do so it needs to absorb the requisite amount of heat to aid it in taking that form, and this heat it must take up from all surrounding warm objects. It absorbs quickly all it can get out of itself as liquid water, out of the glass vessel containing it, and from the surrounding air. But the process of gasification with ebullition goes on so quickly that the temperature of the water thus robbed of heat quickly falls to 0° C., and the remaining water freezes. Thus, then, by pumping out the air from a vessel, *i.e.* working in a vacuum, we can boil a liquid in such exhausted vessel far below its ordinary boiling temperature [Pg 32] in the open air. This fact is of the utmost industrial importance. But touching this question of latent heat, you may ask me for my proof that there is latent heat, and a large amount of it, in a substance that feels perfectly cold. I have told you that a gasified liquid, or a liquefied solid, or most of all a gasified solid, contains such heat, and if reconverted into liquid and solid forms respectively, that heat is evolved, or becomes sensible heat, and then it can be decidedly felt and indicated by the thermometer. Take the case of a liquid suddenly solidifying. The heat latent in that liquid, and necessary to keep it a liquid, is no longer necessary and comes out, and the substance appears to become hot. Quicklime is a cold, white, solid substance, but there is a compound of water and lime—slaked lime—which is also a solid powdery substance, called by the chemist, hydrate of lime. The water used to slake the quicklime is a liquid, and it may be ice-cold water, but to form hydrate of lime it must assume a solid form, and hence can and does dispense with its heat of liquefaction in the change of state. You all know how hot lime becomes on slaking with water. Of course we have heat of chemical combination here as well as evolution of latent heat. As another example, we may take a solution of acetate of soda, so strong that it is just on the point of crystallising. If it crystallises it solidifies, and the liquid consequently gives up its latent heat of liquefaction. We will make it crystallise, first connecting the tube containing it to another one containing a coloured liquid and closed by a cork carrying a narrow tube dipping into the coloured liquid. On crystallising, the solution gives off heat, as is shown by the expansion of the air in the corked tube, and the consequent forcing of the coloured liquid up the narrow tube. Consequently in your works you never dissolve a salt or crystal in water or other liquid without rendering heat latent, or consuming heat; you never allow steam to condense in the steam pipes about the premises without [Pg 33] losing vastly more heat than possibly many are aware of. Let us inquire as to the latent heat of water and of steam.

Latent Heats of Water and Steam.—If we mix 1 kilogram (about 2 lb.) of ice (of course at zero or 0° C.) with 1 kilogram of water at 79° C., and stir well till the ice is melted, *i.e.* has changed its state from solid to liquid, we find, on putting a thermometer in, the temperature is only 0° C. This simply means that 79° of heat (centigrade degrees) have become latent, and represent the heat of liquefaction of 1 kilogram of ice. Had we mixed 1 kilogram of water at 0° C. with 1 kilogram of water at 79° C. there would have been no change of state, and the temperature of the mixture might be represented as a distribution of the 79° C. through the whole mass of the 2 kilograms, and so would be 39½° C. We say, therefore, the latent heat of water is the heat which is absorbed or rendered latent when a unit of weight, say 1 kilogram of water as ice, melts and liquefies to a unit of water at zero, or it is 79 heat units. These 79 units of heat would raise 79 units of weight of liquid water through 1° C., or one unit of liquid water through 79°.

Let us now inquire what the latent heat of steam is. If we take 1 kilogram of water at 0° C. and blow steam from boiling water at 100° C. into it until the water just boils, and then stop and weigh the resulting water, we shall find it amounts to 1·187 kilograms, so that 0·187 kilogram of water which was in the gaseous steam form, and had besides a sensible heat of 100° C., has changed its state to that of liquid water. This liquid water, being at the boiling-point, has still the 100° C. of sensible heat, and hence the water in the gaseous steam form can have given up to the water at 0° C. into which it was blown, only the latent heat of gasification which was not sensible, but by virtue of which it was enabled to assume the gaseous form. But if 0·187 kilogram of steam at 100° C. can heat 1 kilogram of water through [Pg 34] 100 degrees, then 1 kilogram of steam can raise 5·36 kilograms of ice-cold water through 100 degrees, or 536 kilograms through 1 degree, and thus the latent heat of steam is 536 heat units.

Effect of Increase of Pressure on the Boiling of Water.—Now we have referred to diminution of pressure and its effect on the boiling-point of water, and I may point out that by increasing the pressure, such, *e.g.*, as boiling water under a high pressure of steam, you raise the boiling-point. There are some industrial operations in which the action of certain boiling solutions is unavailing to effect certain decompositions or other ends when the boiling is carried on under the ordinary atmospheric pressure, and boiling in closed and strong vessels under pressure must be resorted to. Take as an example the wood-pulp process for making paper from wood shavings. Boiling in open pans with caustic soda lye is insufficient to reduce the wood to pulp, and so boiling in strong vessels under pressure is adopted. The temperature of the solution rises far above 212° F. (100° C.). Let us see what may result chemically from the attainment of such high temperatures of water in our steam boilers working under high pressures. If you blow ordinary steam at 212° F. or 100° C., into fats or oils, the fats and oils remain undecomposed; but suppose you let fatty and oily matters of animal or vegetable origin, such as lubricants, get into your boiler feed-water and so into your boiler, what will happen? I have only to tell you that a process is patented for decomposing fats with superheated steam, to drive or distil over the admixed fatty acids and glycerin, in order to show you that in your boilers such greasy matters will be more or less decomposed. Fats are neutral as fats, and will not injure the iron of the boilers; but once decompose them and they are split up into an acid called a fat acid, and glycerin. That fat acid at the high temperature soon attacks your boilers and pipes, and eats away the iron. That is one of the curious results that may follow at such high temperatures. Mineral or hydrocarbon [Pg 35] oils do not contain these fat acids, and so cannot possibly, even with high-pressure steam, corrode the boiler metal.

Effect of Dissolved Salts on the Boiling of Water.—Let us inquire what this effect is? Suppose we dissolve a quantity of a salt in water, and then blow steam at 100° C. (212° F.) into that water, the latter will boil not at 212° F., but at a higher temperature. There is a certain industrial process I know of, in course of which it is necessary first to maintain a vessel containing water, by means of a heated closed steam coil, at 212° F. (100° C.), and at a certain stage to raise the temperature to about 327° F. (164° C.). The pressure on the boiler connected with the steam coil is raised to nearly seven atmospheres, and thus the heat of the high-pressure steam rises to 327° F. (164° C.), and then a considerable quantity of nitrate of ammonium, a crystallised salt, is thrown into the water, in which it dissolves. Strange to say, although the water alone would boil at 212° F., a strong solution in water of the ammonium nitrate only boils at 327° F., so that the effect of dissolving that salt in the water is the same as if the pressure were raised to seven atmospheres. Now let us, as hat manufacturers, learn a practical lesson from this fact. We have observed that wool and fur fibres are injured by boiling in pure water, and the heat has much to do with this damage; but if the boiling take place in bichrome liquors or similar solutions, that boiling will, according to the strength of the solution in dissolved matters, take place at a temperature more or less elevated above the boiling-point of water, and so the damage done will be the more serious the more concentrated the liquors are, quite independently of the nature of the substances dissolved in those liquors.

Solution. — We have already seen that when a salt of any kind dissolves in water, heat is absorbed, and becomes latent; in other words, cold is produced. I will describe a remarkable example or experiment, well illustrating this fact. If you take [Pg 36] some Glauber's salt, crystallised sulphate of soda, and mix it with some hydrochloric acid (or spirits of salt), then so rapidly will the solution proceed, and consequently so great will be the demand for heat, that if a vessel containing water be put in amongst the dissolving salt, the heat residing in that vessel and its water will be rapidly extracted, and the water will freeze. As regards solubility, some salts and substances are much more quickly and easily dissolved than others. We are generally accustomed to think that to dissolve a substance quickly we cannot do better than build a fire under the containing vessel, and heat the liquid. This is often the correct method of proceeding, but not always. Thus it would mean simply loss of fuel, and so waste of heat, to do this in dissolving ordinary table salt or rock salt in water, for salt is as soluble in cold water as in hot. Some salts are, incredible though it may appear, less soluble in boiling water than in cold. Water just above the freezing-point dissolves nearly twice as much lime as it does when boiling. You see, then, that a knowledge of certain important facts like these may be so used as to considerably mitigate your coal bills, under given circumstances and conditions.

[Pg 37]

LECTURE IV

WATER: ITS CHEMISTRY AND PROPERTIES; IMPURITIES AND THEIR ACTION; TESTS OF PURITY—*Continued*

In the last lecture, under the head of "Solution," I mentioned that some salts, some chemical substances, are more soluble in water than others, and that their solubilities under different circumstances of temperature vary in different ways. However, some salts and compounds are practically insoluble in water under any circumstances. We now arrive at the important result known to chemists as the precipitation of insoluble compounds from solutions. In order to define this result, however, we must, of course, first consider the circumstances of causation of the result. Let us take a simple case of chemical decomposition resulting in the deposition or precipitation of a substance from solution in the insoluble state. We will take a salt you are probably acquainted with—sulphate of copper, or bluestone, and dissolve it in water, and we have then the sulphate of copper in solution in water. Now suppose it is our desire to obtain from that solution all the copper by depositing it in some insoluble form. We may accomplish this in several different ways, relying on certain methods of decomposing that sulphate of copper. One of the simplest and most economical is that adopted in a certain so-called wet method of extracting copper. It is based on the fact that metallic iron has a greater tendency to combine in water solutions, with the acids of copper salts, than the copper [Pg 38] has in those salts. We simply need to place some scraps of iron in the copper sulphate solution to induce a change which may be represented as follows: Copper sulphate, consisting of a combination of copper oxide with sulphuric acid, yields with iron, iron sulphate, a combination of iron oxide with sulphuric acid, and metallic copper. The metallic copper produced separates in the form of a red coating on the iron scraps. But we may also, relying on the fact that oxide of copper is insoluble in water, arrange for the deposition of the copper in that form. This we can do by adding caustic soda to a hot solution of copper sulphate, when we get the following change: Copper sulphate, consisting of a combination of copper oxide with sulphuric acid, yields with caustic soda, sulphate of soda, a combination of soda with sulphuric acid and oxide of copper. Oxide of copper is black, and so in this decomposition what is called a "black precipitate" of that oxide is produced on adding the caustic soda. But it might not suit us thus to deposit the copper from our solution; we might desire to remove the sulphuric acid from the copper sulphate, and leave the copper dissolved, say in the form of a chloride. We select, then, a compound which is a chloride, and a chloride of a metal which forms an insoluble combination with sulphuric acid—chloride of barium, say. On adding this chloride of barium to sulphate of copper solution, we get then a change which we might represent thus: Copper sulphate, consisting of a combination of copper oxide with sulphuric acid, yields with barium chloride, which is a combination of barium and chlorine, insoluble barium sulphate, a combination of barium oxide with sulphuric acid, and soluble copper chloride, a combination of copper and chlorine. This is called a double interchange. Now these are a few illustrations to show you what is meant by chemical decompositions. One practical lesson, of course, we may draw is this: We must have a care in dissolving bluestone or copper sulphate, not to attempt it in iron pans, and not to store or put verdigris into [Pg 39] iron vessels, or the iron will be acted upon, and to some extent the copper salt will become contaminated with iron. It will now be clear to you that, as a solvent for bodies usually soluble in water, water that is perfectly pure will be most suitable and not likely to cause any deposition or precipitation through chemical decompositions, for there are no salts or other compounds in pure water to cause such changes. Such pure water is called soft water. But the term is only a comparative one, and water that is not quite, but nearly pure—pure enough for most practical purposes—is also called soft water. Now rain is the purest form of natural water, for it is a kind of distilled water. Water rises in vapour from the ocean as from a still, and the salt and other dissolved matters remain behind. Meeting cold currents of air, the vapours condense in rain, and fall upon the earth. After coming in contact with the earth, the subsequent condition of that water entirely depends upon the character, as regards solubility or insolubility, of the substances composing the strata or layers of earth upon which it falls, and through which it sinks. If it meets with insoluble rocks—for all rocks are not insoluble—it remains, of course, pure and soft, and in proportion as the constituents of rock and soil are soluble, in that proportion does the water become hard. We all know how dangerous acid is in water, causing that water to act on many substances, the iron of iron vessels, the lime in soil or rock, etc., bringing iron and lime respectively into solution. Now the atmosphere contains carbonic acid, and carbonic acid occurs in the earth, being evolved by decomposing vegetation, etc. Carbonic acid is also soluble to a certain, though not large extent, in water. As we shall see, water charged with carbonic acid attacks certain substances insoluble in pure water, and brings them into solution, and thus the water soon becomes hard. About the close of the last lecture, I said that lime is, to a certain extent, soluble in cold water. The solution is called lime-water; it might [Pg 40] be called a solution of caustic lime. When carbonic acid gas first comes in contact with such a solution, chalk or carbonate of lime, which is insoluble in water, is formed, and the lime is thus precipitated as carbonate. Supposing, however, we continued to pass carbonic acid gas into that water, rendered milky with chalk powder, very soon the liquid would clear, and we should get once more a solution of lime, but not caustic lime as it was at first, simply now a solution of carbonate of lime in carbonic acid, or a solution of bicarbonate of lime. I will take some lime-water, and I will blow through; my breath contains carbonic acid, and you will see the clear liquid become milky owing to separation of insoluble carbonate of lime, or chalk. I now continue blowing, and at length that chalk dissolves with the excess of carbonic acid, forming bicarbonate of lime. This experiment explains how it is that water percolating through or running over limestone strata dissolves out the insoluble chalk. Such water, hard from dissolved carbonate of lime, can be softened by merely boiling the water, for the excess of carbonic acid is then expelled, and the chalk is precipitated again. This would be too costly for the softening of large quantities of water, the boiling process consuming too much coal, and so another process is adopted. Quicklime, or milk of lime, is added to the water in the proper quantity. This lime unites with the excess of carbonic acid holding chalk in solution, and forms with it insoluble chalk, and so all deposits together as chalk. By this liming process, also, the iron of the water dissolved likewise in ferruginous streams, etc., by carbonic acid, would be precipitated. To show this deposition I will now add some clear lime-water to the solution I made of chalk with the carbonic acid of my breath, and a precipitate is at once formed, all the lime and carbonic acid together depositing as insoluble chalk. Hence clear lime-water forms a good test for the presence of bicarbonates of lime or [Pg 41] iron in a water. But water may be hard from the presence of other salts, other lime salts. For example, certain parts of the earth contain a great deal of gypsum, or natural sulphate of lime, and this is soluble to some extent in water. Water thus hardened is not affected by boiling, or the addition of lime, and is therefore termed permanently hard water, the water hardened with dissolved chalk being termed temporarily hard water. I have said nothing of solid or undissolved impurities in water, which are said to be in suspension, for the separation of these is a merely mechanical matter of settling, or filtration and settling combined. As a general rule, the water of rivers contains the most suspended and vegetable matter and the least amount of dissolved constituents, whereas spring and well waters contain the most dissolved matters and the least suspended. Serious damage may be done to the dyer by either of these classes of impurities, and I may tell you that the dissolved calcareous and magnesian impurities are the most frequent in occurrence and the most injurious. I told you that on boiling, the excess of carbonic acid holding chalk or carbonate of lime in solution as bicarbonate, is decomposed and carbonate of lime precipitated. You can at once imagine, then, what takes place in your steam boilers when such water is used, and how incrustations are formed. Let us now inquire as to the precise nature of the waste and injury caused by hard and impure waters. Let us also take, as an example, those most commonly occurring injurious constituents, the magnesian and calcareous impurities. Hard water only produces a lather with soap when that soap has effected the softening of the water, and not till then. In that process the soap is entirely wasted, and the fatty acids in it form, with the lime and magnesia, insoluble compounds called lime and magnesia soaps, which are sticky, greasy, adhesive bodies, that precipitate and fix some colouring matters like a mordant. We have in such cases, then, a kind of double mischief—(i) waste of soap, (ii) injury to colours and [Pg 42] dyes on the fabrics. But this is not all, for colours are precipitated as lakes, and mordants also are precipitated, and thus wasted, in much the same sense as the soaps are. Now by taking a soap solution, formed by dissolving a known weight of soap in a known volume of water, and adding this gradually to hard water until a permanent lather is just produced, we can directly determine the consumption of soap by such a water, and ascertain the hardness. Such a method is called Clark's process of determination or testing, or Clark's soap test. We hear a great deal just now of soaps that will wash well in hard water, and do wonders under any conditions; but mark this fact, none of them will begin to perform effective duty until such hard water has been rendered soft at the expense of the soap. Soaps made of some oils, such as cocoa-nut oil, for example, are more soluble in water than when made of tallow, etc., and so they more quickly soften a hard water and yield lather, but they are wasted, as far as consumption is concerned, to just the same extent as any other soaps. They do not, however, waste so much time and trouble in effecting the end in view, and, as you know, "Time is money" in these days of work and competition. After making a soap test as described above, and knowing the quantity of water used, it is, of course, easy to calculate the annual loss of soap caused by the hardness of the water. The monthly consumption of soap in London is 1,000,000 kilograms (about 1000 tons), and it is estimated that the hardness of the Thames water means the use of 230,000 kilograms (nearly 230 tons) more soap per month than would be necessary if soft water were used. Of course the soap manufacturers around London would not state that fact on their advertising placards, but rather dwell on the victorious onslaught their particular brand will make on the dirt in articles to be washed, in the teeth of circumstances that would be hopeless for any other brand of soap! I have referred to the sticky and adhesive character of the compounds called lime soaps, formed [Pg 43] in hard waters. Now in washing and scouring wool and other fibres, these sticky lime soaps adhere so pertinaciously that the fibres, be they of wool, silk, or any other article, remain in part untouched, impermeable to mordant or colouring matter, and hence irregular development of colour must be the consequence. Also an unnatural lustre or peculiar bloom may in parts arise, ruining the appearance of the goods. In some cases the lime soaps act like mordants, attracting colouring matter unequally, and producing patchy effects. In the dye-baths in which catechu and tannin are used, there is a waste of these matters, for insoluble compounds are formed with the lime, and the catechu and tannin are, to a certain extent, precipitated and lost. Some colours are best developed in an acid bath, such as Cochineal Scarlet, but the presence of the bicarbonate of lime tends to cause neutralisation of the acidity, and so the dyeing is either retarded or prevented. Such mordants as "red liquor" and "iron liquor," which are acetates of alumina and iron respectively, are also wasted, a portion of them being precipitated by the lime, thus weakening the mordant baths.

Ferruginous Impurities in Water. — Iron in solution in water is very objectionable in dyeing operations. When the iron is present as bicarbonate, it acts on soap solutions like the analogous lime and magnesia compounds, producing even worse results. In wool scouring, cotton bleaching, and other processes requiring the use of alkaline carbonates, ferric oxide is precipitated on the fibre. A yellowish tinge is communicated to bleached fabrics, and to dye bright and light colours is rendered almost out of the question. You may always suspect iron to be present in water flowing from or obtained directly out of old coal pits, iron mines, or from places abounding in iron and aluminous shales. Moreover, you sometimes, or rather generally, find that surface water draining off moorland districts, and passing over ochre beds, contains iron, and on its way deposits on the beds of the streamlets conveying it, and on the [Pg 44] stones, red or brown oxide of iron. All water of this kind ought to be avoided in dyeing and similar operations. The iron in water from old coal pits and shale deposits is usually present as sulphate due to the oxidation of pyrites, a sulphuret or sulphide of iron. Water from heaths and moorlands is often acid from certain vegetable acids termed "peaty acids." This acidity places the water in the condition of a direct solvent for iron, and that dissolved iron may cause great injury. If such water cannot be dispensed with, the best way is to carefully neutralise it with carbonate of soda; the iron is then precipitated as carbonate of iron, and can be removed.

Contamination of Water by Factories. — You may have neighbours higher up the stream than yourselves, and these firms may cast forth as waste products substances which will cause immense waste and loss. Amongst these waste products the worst are those coming from chemical works, paper works, bleach works, etc. If the paper works be those working up wood pulp, the pollutions of effluent water will be about as noxious as they well can be. You will have gums and resins from the wood, calcium chloride from the bleach vats, acids from the "sours"; resin, and resin-soaps; there may also be alumina salts present. Now alumina, lime, resin, and resin-soaps, etc., precipitate dyestuffs, and also soap; if the water is alkaline, some of the mordants used may be precipitated and wasted, and very considerable damage done.

Permanent hardness in water, due to the presence of gypsum or sulphate of lime in solution, may be remedied by addition of caustic soda. Of course, if an alkaline water is objectionable in any process, the alkali would have to be neutralised by the addition of some acid. For use in boilers, water might thus be treated, but it would become costly if large quantities required such treatment. Water rendered impure by contaminations from dyehouses and some chemical works can be best purified, and most cheaply, by simple liming, followed [Pg 45] by a settling process. If space is limited and much water is required, instead of the settling reservoirs, filtering beds of coke, sand, etc., may be used. The lime used neutralises acids in the contaminated and impure water, precipitates colouring matters, mordants, soap, albuminous matters, etc.

Tests of Purity.—I will now describe a few tests that may be of value to you in deciding as to what substances are contaminating any impure waters that may be at hand.

Iron.—If to a water you suspect to be hard from presence of carbonate of lime or carbonate of iron in solution in carbonic acid, *i.e.* as bicarbonates, you add some clear lime-water, and a white precipitate is produced, you have a proof of carbonate of lime—hardness. If the precipitate is brownish, you may have, also, carbonate of iron. I will now mention a very delicate test for iron. Such a test would be useful in confirmation. If a very dilute solution of such iron water be treated with a drop or two of pure hydrochloric acid, and a drop or so of permanganate of potash solution or of Condy's fluid, and after that a few drops of yellow prussiate of potash solution be added, then a blue colour (Prussian blue), either at once or after standing a few hours, proves the presence of iron.

Copper.—Sometimes, as in the neighbourhood of copper mines or of some copper pyrites deposits, a water may be contaminated with small quantities of copper. The yellow prussiate once more forms a good test, but to ensure the absence of free mineral acids, it is first well to add a little acetate of soda solution. A drop or two of the prussiate solution then gives a brown colour, even if but traces of copper are present.

Magnesia.—Suppose lime and magnesia are present. You may first evaporate to a small bulk, adding a drop of hydrochloric acid if the liquid becomes muddy. Then add ammonia and ammonium oxalate, when lime alone is precipitated as the oxalate of lime. Filter through blotting paper, and to the [Pg 46] clear filtrate add some phosphate of soda solution. A second precipitation proves the presence of magnesia.

Sulphates.—A solution of barium chloride and dilute hydrochloric acid gives a white turbidity.

Chlorides.—A solution of silver nitrate and nitric acid gives a white curdy precipitate.

Test for Lead in Drinking Water.—I will, lastly, give you a test that will be useful in your own homes to detect minute quantities of lead in water running through lead pipes. Place a large quantity of the water in a glass on a piece of white paper, and add a solution of sulphuretted hydrogen and let stand for some time. A brown colour denotes lead. Of course copper would also yield a brown coloration, but I am supposing that the circumstances preclude the presence of copper.

I have already said that rain water is the purest of natural waters; it is so soft, and free from dissolved mineral matters because it is a distilled water. In distilling water to purify it, we must be very careful what material we use for condensing the steam in, since it is a fact probably not sufficiently well known, that the softer and purer a water is, the more liable it is to attack lead pipes. Hence a coil of lead pipe to serve as condensing worm would be inadmissible. Such water as Manchester water, and Glasgow water from Loch Katrine still more so, are more liable to attack lead pipes than the hard London waters. To illustrate this fact, we will distil some water and condense in a leaden worm, then, on testing the water with our reagent, the sulphuretted hydrogen water, a brown colour is produced, showing the presence of lead. On condensing in a block tin worm, however, no tin is dissolved, so tin is safer and better as the material for such a purpose than lead.

Filtration.—We hear a great deal about filtration or filters as universal means of purifying water. Filtration, we must remember, will, as a rule, only remove solid or suspended impurities in water. For example, if we take some ivory black [Pg 47] or bone black, and mix it with water and afterwards filter the black liquid through blotting-paper, the bone black remains on the paper, and clear, pure water comes through. Filtering is effective here. If we take some indigo solution, however, and pour it on to the filter, the liquid runs through as blue as it was when poured upon the filter. Filtering is ineffective here, and is so generally with liquids containing matters dissolved in them. But I said "generally," and so the question is suggested—Will filtration of any kind remove matters in solution? This question I will, in conclusion, try to answer. Bone charcoal, or bone black, has a wonderful attraction for many organic matters such as colours, dyes, and coloured impurities like those in peat water, raw sugar solutions, etc. For example, if we place on a paper filter some bone black, and filter through it some indigo solution, after first warming the latter with some more of the bone black, the liquid comes through clear, all the indigo being absorbed in some peculiar way, difficult to explain, by the bone black, and remaining on the filter. This power of charcoal also extends to gases, and to certain noxious dissolved organic impurities, but it is never safe to rely too much on such filters, since the charcoal can at length become charged with impurities, and gradually cease to act. These filters need cleaning and renewing from time to time.

[Pg 48]

LECTURE V

ACIDS AND ALKALIS

Properties of Acids and Alkalis.—The name acids is given to a class of substances, mostly soluble in water, having an acid or sour taste, and capable of turning blue litmus solution red. All acids contain one or more atoms of hydrogen capable of being replaced by metals, and when such hydrogen atoms are completely replaced by metals, there result so-called neutral or normal salts, that is, neutral substances having no action on litmus solution. These salts can also be produced by the union of acids with equivalent quantities of certain metallic oxides or hydroxides, called bases, of which those soluble in water are termed alkalis. Alkalis have a caustic taste, and turn red litmus solution blue.

In order to explain what is called the law of equivalence, I will remind you of the experiment of the previous lecture, when a piece of bright iron, being placed in a solution of copper sulphate, became coated with metallic copper, an equivalent weight of iron meanwhile suffering solution as sulphate of iron. According to the same law, a certain weight of soda would always require a certain specific equivalent weight of an acid, say hydrochloric acid, to neutralise its alkaline or basic properties, producing a salt.

The specific gravities of acids and alkalis in solution are made use of in works, etc., as a means of ascertaining their strengths and commercial values. Tables have been carefully [Pg 49] constructed, such that for every degree of specific gravity a corresponding percentage strength of acidity and alkalinity may be looked up. The best tables for this purpose are given in Lunge and Hurter's *Alkali-Makers' Pocket-Book*, but for ordinary purposes of calculation in the works or factory, a convenient relationship exists in the case of hydrochloric acid between specific gravity and percentage of real acid, such that specific gravity as indicated by Twaddell's hydrometer directly represents percentage of real acid in any sample of hydrochloric acid.

The point at which neutralisation of an acid by alkali or *vice versâ* just takes place is ascertained very accurately by the use of certain sensitive colours. At first litmus and cochineal tinctures were used, but in testing crude alkalis containing alumina and iron, it was found that lakes were formed with these colours, and they become precipitated in the solution, and so no longer sensitive. The chemist was then obliged to resort to certain sensitive coal-tar colours, which did not, as the dyer and printer knew, form lakes with alumina and iron, such as methyl orange, fluorescein, Congo red, phenolphthalein, and so forth. For determining the alkalimetric strength of commercial sodas, a known weight of the sample is dissolved in water, and a few drops of a solution of methyl orange are added, which colour the solution yellow or orange. Into this solution is then run, from a burette or graduated tube, a standard solution of an acid, that is, a solution prepared by dissolving a known weight of an acid, say hydrochloric acid, in a known volume of water. The acid is run in gradually until the yellow colour changes to pink, at which point the volume of acid used is noted. Knowing the weight of acid contained in this volume of standard acid, and having regard to the law of equivalence mentioned above, it is an easy matter to calculate the amount of alkali equivalent to the acid used, and from this the alkali contained in the sample.

[Pg 50]

Sulphuric Acid.—The first process for manufacturing sulphuric acid or vitriol was by placing some burning sulphur in a closed vessel containing some water. The water absorbed the acid formed by the burning sulphur. It was next discovered that by mixing with the sulphur some nitre, much more sulphuric acid could be produced per given quantity of brimstone. At first large glass carboys were used, but in 1746 the carboys were replaced by chambers of lead containing water at the bottom, and in these lead chambers the mixture of sulphur and nitre was burnt on iron trays. Next, although gradually, the plant was divided into two portions—a furnace for burning the sulphur, and a chamber for receiving the vapours. The system was thus developed into the one followed at the present time. The sulphur, or, in most cases, cupreous iron pyrites (a combination of iron and copper with sulphur), is burned in specially constructed kilns or furnaces, and the hot gases, consisting essentially of sulphur dioxide with the excess of air, pass through flues in which are placed cast-iron "nitre pots" containing a mixture of nitre (sodium nitrate) and vitriol. The gases thus become mixed with nitrous fumes or gaseous oxides of nitrogen, and, after cooling, are ready for mixing with steam or water spray in the lead chambers in which the vitriol is produced. These oxides of nitrogen enable the formation of sulphuric acid to take place more quickly by playing the part of oxygen-carriers. Sulphuric acid is formed by the union of oxygen with sulphur dioxide and water; the oxides of nitrogen combine with the oxygen of the air present in the chambers, then give up this oxygen to the sulphur dioxide and water or steam to form sulphuric acid, again combine with more oxygen, and so on. The exact processes or reactions are of course much more complicated, but the above represents what is practically the ultimate result. It is evident that the gases leaving the last lead chamber in which the formation of vitriol is effected, must still contain nitrous fumes, [Pg 51] and it becomes a matter of importance to recover them, so that they can be used over again. To effect this object, use is made of the solubility of nitrous fumes in strong vitriol. The gases from the last lead chamber of the series are passed through what is called a Gay-Lussac tower (the process was invented by the eminent French chemist Gay-Lussac), which is a tower made of lead, supported by a wooden framework, and filled with coke or special stoneware packing, over which strong vitriol is caused to flow. The vitriol dissolves the nitrogen oxides, and so-called "nitrous vitriol" flows out at the base of the tower. The recovery of the nitrogen compounds from the nitrous vitriol is effected in Glover towers (the invention of John Glover of Newcastle), which also serve to concentrate to some extent the weak acid produced in the lead chambers, and to cool the hot gases from the sulphur burners or pyrites kilns. The weak chamber acid is mixed with the nitrous vitriol from the Gay-Lussac tower, and the mixture is pumped to the top of the Glover tower, which is of similar construction to the Gay-Lussac tower, but is generally packed with flints. This Glover tower is placed between the sulphur burners or pyrites kilns and the first lead chamber. The nitrous vitriol passing down the tower meets the hot gases from the kilns, and a threefold object is effected: (1) The nitrous fumes are expelled from the nitrous vitriol, and are carried into the chambers, to again play the part of oxygen-carriers; (2) the weak chamber acid which was mixed with the nitrous vitriol is concentrated by the hot kiln gases; and (3) the hot gases themselves are cooled. The acid from the Glover tower is purified by special treatment—for example, the arsenic may be removed, after precipitation with sulphuretted hydrogen, in the form of insoluble arsenic sulphide,—and the purified acid is concentrated by heating in glass or platinum vessels.

A considerable amount of sulphuric acid is now made by the so-called "contact process," in which sulphur dioxide and [Pg 52] oxygen unite to form sulphuric acid in presence of a heated "contact" substance, usually some form of finely-divided platinum.

Nitric Acid.—This acid is usually prepared by distilling a mixture of sodium nitrate and vitriol in cast-iron retorts or pots, the nitric acid being collected in stoneware vessels connected one with another, or, as is more generally the case at the present time, in condensing apparatus consisting of stoneware pipes or coils cooled by water. The effluent gases are passed through a scrubber in order to free them from the last traces of acid before discharging them into the atmosphere.

Hydrochloric Acid.—The greater part of the hydrochloric acid manufactured in Great Britain is obtained as an intermediate product in the Leblanc alkali process, which will presently be described, being produced by heating common salt with vitriol. A large quantity is, however, also produced by the so-called direct process of Hargreaves & Robinson, which is, in principle, the same method as that employed in the Leblanc process, except that the intermediate product, vitriol, is not separated. It consists essentially in passing the hot gases from pyrites kilns, as used in the manufacture of vitriol, through large cast-iron vessels containing common salt heated to a high temperature. Various physical conditions must be complied with in order to make the process a success. For example, the salt is used in the form of moulded hard porous cakes made from a damp mixture of common salt and rock salt. The cast-iron vessels must be heated uniformly, and the hot pyrites kiln gases must be passed downwards through the salt in order to ensure uniform distribution. The hydrochloric acid is condensed in stoneware pipes connected with towers packed with coke or stoneware.

Alkali: Leblanc Process.—The manufacture of vitriol, as I have described it to you, is the first step in the Leblanc process. The next stage consists in the manufacture of sodium [Pg 53] sulphate (salt-cake) and hydrochloric acid from the sulphuric acid and common salt; this is called the salt-cake process. The production of salt-cake or crude sodium sulphate is carried out in two stages. A large covered iron pan, called the decomposing pan or salt-cake pot, is mounted in one part of the salt-cake furnace, and alongside it is the hearth or bed on which the second stage of the process, the drying or roasting, is effected. The mixture of common salt and vitriol is charged into the salt-cake pot, which is heated by a fire below. When from two-thirds to three-quarters of the hydrochloric acid has been expelled from the charge, the mass acquires the consistence of thick dough, and at this stage it is raked out of the pan on to the roasting hearth alongside, where the decomposition is completed by means of flames playing directly on to the top of the charge. The hydrochloric acid evolved during the process is condensed in much the same manner as in the process of Hargreaves & Robinson previously described. It is a curious fact that in the earlier years of the Leblanc process, hydrochloric acid, or "spirits of salt," as it is frequently called, was a by-product that required all the vigilance of the alkali-works inspectors to prevent it being allowed to escape from the chimneys in more than a certain small regulated amount. Now, it is the principal product; indeed, the Leblanc alkali maker may be said to subsist on that hydrochloric acid, as his chief instrument for producing chloride of lime or bleaching powder.

Mechanical furnaces are now used to a large extent for the salt-cake process. They consist broadly of a large revolving furnace-hearth or bed, on to which the mixture of salt and vitriol is charged, and on which it is continuously agitated, and gradually moved to the place of discharge, by rakes or the like, operated by suitable machinery.

The next stage of the Leblanc process is the manufacture of "black ash," or crude sodium carbonate. This is usually done [Pg 54] in large cylindrical revolving furnaces, through, which flames from a fire-grate, or from the burning of gaseous fuel, pass; the waste heat is utilised for boiling down "black ash" liquor, obtained by lixiviating the black ash. A mixture of salt-cake, limestone or chalk (calcium carbonate), and powdered coal or coal slack is charged into the revolving cylinder; during the process the mass becomes agglomerated, and the final product is what is known as a "black-ash ball," consisting chiefly of crude sodium carbonate and calcium sulphide, but containing smaller quantities of many other substances. The soda ash or sodium carbonate is obtained from the black ash by lixiviating with water, and after various purification processes, the solution is boiled down, as previously stated, by the waste heat of the black-ash furnace. The alkali is sold in various forms as soda ash, soda crystals, washing soda, etc.

Caustic soda is manufactured from solution of carbonate of soda by causticising, that is, treatment with caustic lime or quicklime.

It will have been noticed that one of the chief reagents in the Leblanc process is the sulphur used in the form of brimstone or as pyrites for making vitriol in the first stage; this sulphur goes through the entire process; from the vitriol it goes to form a constituent of the salt-cake, and afterwards of the calcium sulphide contained in the black ash. This calcium sulphide remains as an insoluble mass when the carbonate of soda is extracted from the black ash, and forms the chief constituent of the alkali waste, which until the year 1880 could be seen in large heaps around chemical works. Now, however, by means of treatment with kiln gases containing carbonic acid, the sulphur is extracted from the waste in the form of hydrogen sulphide, which is burnt to form vitriol, or is used for making pure sulphur; and so what was once waste is now a source of profit.

Ammonia-Soda Process of Alkali Manufacture. — This process [Pg 55] depends upon the fact that when carbonic acid is forced, under pressure, into a saturated solution of ammonia and common salt, sodium bicarbonate is precipitated, whilst ammonium chloride or "sal-ammoniac" remains dissolved in the solution. The reaction was discovered in 1836 by a Scotch chemist named John Thom, and small quantities of ammonia-soda were made at that time by the firm of McNaughton & Thom. The successful carrying out of the process on the large scale depends principally upon the complete recovery of the expensive reagent, ammonia, and this problem was only solved within comparatively recent years by Solvay. The process has been perfected and worked with great success in England by Messrs. Brunner, Mond, & Co., and has proved a successful rival to the Leblanc process.

Alkali is also produced to some extent by electrolytic processes, depending upon the splitting up of a solution of common salt into caustic soda and chlorine by the use of an electric current.

[Pg 56]

LECTURE VI

BORIC ACID, BORAX, SOAP

Boric Acid. — At ordinary temperatures and under ordinary conditions boric acid is a very weak acid, but like silicic and some other acids, its relative powers of affinity and combination become very much changed at high temperatures; thus, fused and strongly heated boric acid can decompose carbonates and even sulphates, and yet a current of so weak an acid as hydrogen sulphide, passed through a strong solution of borax, will decompose it and set free boric acid. Boric acid is obtained chiefly from Italy. In a tract of country called the Maremma of Tuscany, embracing an area of about forty square miles, are numerous chasms and crevices, from which hot vapour and heated gases and springs of water spurt. The steam issuing from these hot springs contains small quantities of boric acid, that acid being one of those solid substances distilling to some extent in a current of steam. The steam vapours thus bursting forth, owing to some kind of constant volcanic disturbance, are also more or less laden with sulphuretted hydrogen gas, communicating a very ill odour to the neighbourhood. These phenomena were at first looked upon by the people as the work of the devil, and priestly exorcisms were in considerable request in the hope of quelling them, very much as a great deal of the mere speech-making at the present time in England on foreign competition and its evils, and the dulness of trade, the artificial combinations to keep up prices, to reduce wages, general [Pg 57] lamentation, etc., are essayed in the attempt to charm away bad trade. At length a kind of prophet arose of a very practical character in the form of the late Count Lardarel, who, mindful of the fact that the chemist Höffer, in the time of the Grand Duke Leopold I., had discovered boric acid in the volcanic steam jets, looked hopefully beyond the exorcisms of the priests and the superstitions of the people to a possible blessing contained in what appeared to be an unholy confusion of Nature. He constructed tanks of from 100 to 1000 ft. in diameter and 7 to 20 ft. in depth, of such a kind that the steam jets were surrounded by or contained in them, and thus the liquors formed by condensation became more and more concentrated. These tanks were arranged at different levels, so that the liquors could be run off from one to the other, and finally to settling cisterns. Subsequently the strong liquors were run to lead-lined, wooden vats, in which the boric acid was crystallised out. Had the industry depended on the use of fuel it could never have developed, but Count Lardarel ingeniously utilised the heat of the steam for all the purposes, and neither coal nor wood was required. Where would that Tuscan boric acid industry have been now had merely the lamentations of landowners, fears of the people, and exorcisms of the priests been continued? Instead of being the work of the arch-enemy of mankind, was not it rather an incitement to a somewhat high and difficult step in an upward direction towards the attainment, on a higher platform of knowledge and skill, of a blessing for the whole province of Tuscany? What was true in the history of that industry and its development is every whit as true of the much-lamented slackening of trade through foreign competition or other causes now in this country, and coming home to yourselves in the hat-manufacturing industry. The higher platform to which it was somewhat difficult to step up, but upon which the battle must be fought and the victory won, was one of a higher scientific and technological [Pg 58] education and training. The chemist Höffer made the discovery of boric acid in the vapours, they would no doubt take note; but Höffer went no further; and it needed the man of both educated and practical mind like Count Lardarel to turn the discovery to account and extract the blessing. In like manner it was clear that in our educational schemes for the benefit of the people, there must not only be the scientific investigator of abstract truth, but also the scientific technologist to point the way to the practical realisation of tangible profit. Moreover, and a still more important truth, it is the scientific education of the proprietors and heads we want—educated capital rather than educated workmen.

Borax.—A good deal of the Tuscan boric acid is used in France for the manufacture of borax, which is a sodium salt of boric acid. Borax is also manufactured from boronitrocalcite, a calcium salt of boric acid, which is found in Chili and other parts of South America. The crude boronitrocalcite or "tiza" is boiled with sodium carbonate solution, and, after settling, the borax is obtained by crystallisation.

Borax itself is found in California and Nevada, U.S.A., and also in Peru, Ceylon, China, Persia, and Thibet. The commercial product is obtained from the native borax (known as "tincal") by dissolving in water and allowing the solution to crystallise. The Peruvian borax sometimes contains nitre. For testing the purity of refined borax the following simple tests will usually suffice. A solution of the borax is made containing 1 part of borax to 50 parts of water, and small portions of the solution are tested as follows: *Heavy metals* (*lead*, *copper*, etc.).—On passing sulphuretted hydrogen into the solution, no coloration or precipitate should be produced. *Calcium Salts.*—The solution should not give a precipitate with ammonium oxalate solution. *Carbonates.*—The solution should not effervesce on addition of nitric or hydrochloric acid. *Chlorides.*—No appreciable [Pg 59] precipitate should be produced on addition of silver nitrate solution and nitric acid. *Sulphates.*—No appreciable precipitate should be produced on adding hydrochloric acid and barium chloride. *Iron.*—50 c.c. of the solution should not immediately be coloured blue by 0·5 c.c. of potassium ferrocyanide solution.

Soap.—Soap is a salt in the chemical sense, and this leads to a wider definition of the term "salt" or "saline" compound. Fats and oils, from which soaps are manufactured, are a kind of *quasi* salts, composed of a fatty acid and a chemical constant, if I may use the term, in the shape of base, namely, glycerin. When these fats and oils, often called glycerides, are heated with alkali, soda, a true salt of the fatty acid and soda is formed, and this is the soap, whilst the glycerin remains behind in the "spent soap lye." Now glycerin is soluble in water containing dissolved salt (brine), whilst soap is insoluble, though soluble in pure water. The mixture of soap and glycerin produced from the fat and soda is therefore treated with brine, a process called "cutting the soap." The soap separates out in the solid form as a curdy mass, which can be easily separated. Certain soaps are able to absorb a large quantity of water, and yet appear quite solid, and in purchasing large quantities of soap it is necessary, therefore, to determine the amount of water present. This can be easily done by weighing out ten or twenty grams of the soap, cut in small pieces, into a porcelain dish and heating over a gas flame, whilst keeping the soap continually stirred, until a glass held over the dish no longer becomes blurred by escaping steam. After cooling, the dry soap is weighed, and the loss of weight represents the amount of moisture. I have known cases where soap containing about 83 per cent. of water has been sold at the full market price. Some soaps also contain more alkali than is actually combined with the fatty acids of the soap, and that excess alkali is injurious in washing silks and scouring [Pg 60] wool, and is also not good for the skin. The presence of this free or excess alkali can be at once detected by rubbing a little phenolphthalein solution on to the freshly-cut surface of a piece of soap; if free alkali be present, a red colour will be produced.

[Pg 61]

LECTURE VII

SHELLAC, WOOD SPIRIT, AND THE STIFFENING AND PROOFING PROCESS

Shellac. — The resin tribe, of which shellac is a member, comprises vegetable products of a certain degree of similarity. They are mostly solid, glassy-looking substances insoluble in water, but soluble in alcohol and wood spirit. In many cases the alcoholic solutions show an acid reaction. The resins are partly soluble in alkalis, with formation of a kind of alkali salts which we may call resin-soaps.

Shellac is obtained from the resinous incrustation produced on the bark of the twigs and branches of various tropical trees by the puncture of the female "lac insect" (*Taccardia lacca*). The lac is removed from the twigs by "beating" in water; the woody matter floats to the surface, and the resin sinks to the bottom, and when removed forms what is known as "seed-lac." Formerly, the solution, which contains the colouring matter dissolved from the crude "stick-lac," was evaporated for recovery of the so-called "lac-dye," but the latter is no longer used technically. The seed-lac is bleached by boiling with sodium or potassium carbonate, alum, or borax, and then, if it is not pale enough, is further bleached by exposure to sunlight. It is now dried, melted, and mixed with a certain proportion of rosin or of orpiment (a sulphide of arsenic) according to the purpose for which it is desired. After further operations of melting and straining, the lac is melted and spread [Pg 62] into thin sheets to form ordinary shellac, or is melted and dropped on to a smooth surface to form "button-lac." Ordinary shellac almost invariably contains some rosin, but good button-lac is free from this substance. The presence of 5 per cent. of rosin in shellac can be detected by dissolving in a little alcohol, pouring the solution into water, and drying the fine impalpable powder which separates. This powder is extracted with petroleum spirit, and the solution shaken with water containing a trace of copper acetate. If rosin be present, the petroleum spirit will be coloured emerald-green.

Borax, soda crystals, and ammonia are all used to dissolve shellac, and it may be asked: Which of these is least injurious to wool? and why? How is their action modified by the presence of dilute sulphuric acid in the wool? I would say that soda crystals and ammonia are alkalis, and if used strong, are sure to do a certain amount of injury to the fibre of wool, and more if used hot than cold. Of the two, the ammonia will have the least effect, especially if dilute, but borax is better than either. The influence of a little sulphuric acid in the wool would be in the direction of neutralising some of the ammonia or soda, and shellac, if dissolved in the alkalis, would be to some extent precipitated on the fibre, unless the alkali, soda or ammonia, were present in sufficient excess to neutralise that sulphuric acid and to leave a sufficient balance to keep the shellac in solution. Borax, which is a borate of soda, would be so acted on by the sulphuric acid that some boric acid would be set free, the sulphuric acid robbing some of that borax of its soda. This boric acid would not be nearly so injurious to wool as carbonate of soda or ammonia would.

The best solvent for shellac, however, in the preparation of the stiffening and proofing mixture for hats, is probably wood spirit or methylated spirit. A solution of shellac in wood spirit is indeed used for the spirit-proofing of silk hats, and to some [Pg 63] extent of felt hats, and on the whole the best work, I believe, is done with it. Moreover, borax is not a cheap agent, and being non-volatile it is all left behind in the proofed material, whereas wood spirit or methylated spirit is a volatile liquid, *i.e.* a liquid easily driven off in vapour, and after application to the felt it may be almost all recovered again for re-use. In this way I conceive the use of wood spirit would be both more effective and also cheaper than that of borax, besides being most suitable in the case of any kind of dyes and colours to be subsequently applied to the hats.

Wood Spirit.—Wood spirit, the pure form of which is methyl alcohol, is one of the products of the destructive distillation of wood. The wood is distilled in large iron retorts connected to apparatus for condensing the distillation products. The heating is conducted slowly at first, so that the maximum yield of the valuable products—wood acid (acetic acid) and wood spirit—which distil at a low temperature, is obtained. When the condensed products are allowed to settle, they separate into two distinct layers, the lower one consisting of a thick, very dark tar, whilst the upper one, much larger in quantity, is the crude wood acid (containing also the wood spirit), and is reddish-yellow or reddish-brown in colour. This crude wood acid is distilled, and the wood spirit which distils off first is collected separately from the acetic acid which afterwards comes over. The acid is used for the preparation of alumina and iron mordants (see next lecture), or is neutralised with lime, forming grey acetate of lime, from which, subsequently, pure acetic acid or acetone is prepared. The crude wood spirit is mixed with milk of lime, and after standing for several hours is distilled in a rectifying still. The distillate is diluted with water, run off from any oily impurities which are separated, and re-distilled once or twice after treatment with quicklime.

Stiffening and Proofing Process. — Before proceeding to [Pg 64] discuss the stiffening and proofing of hat forms or "bodies," it will be well to point out that it was in thoroughly grasping the importance of a rational and scientific method of carrying out this process that Continental hat manufacturers had been able to steal a march upon their English rivals in competition as to a special kind of hat which sold well on the Continent. There are, or ought to be, three aims in the process of proofing and stiffening, all the three being of equal importance. These are: first, to waterproof the hat-forms; second, to stiffen them at the same time and by the same process; and the third, the one the importance of which I think English hat manufacturers have frequently overlooked, at least in the past, is to so proof and stiffen the hat-forms as to leave them in a suitable condition for the subsequent dyeing process. In proofing the felt, the fibres become varnished over with a kind of glaze which is insoluble in water, and this varnish or proof is but imperfectly removed from the ends of the fibres on the upper surface of the felt. The consequence is a too slight penetration of the dyestuff into the inner pores of the fibres; indeed, in the logwood black dyeing of such proofed felt a great deal of the colour becomes precipitated on the outside of the fibres—a kind of process of "smudging-on" of a black pigment taking place. The subsequent "greening" of the black hats after a short period of wear is simply due to the ease with which such badly fixed dye rubs off, washes off, or wears off, the brownish or yellowish substratum which gradually comes to light, causing a greenish shade to at length appear. If we examine under the microscope a pure unproofed fur fibre, its characteristic structure is quite visible. Examination of an unproofed fibre dyed with logwood black shows again the same characteristic structure with the dye inside the fibre, colouring it a beautiful bluish-grey tint, the inner cellular markings being black. A proofed fur fibre, on the other hand, when examined under the microscope, is seen to be covered with a [Pg 65] kind of translucent glaze, which completely envelops it, and prevents the beautiful markings showing the scaly structure of the fibre from being seen. Finally, if we examine microscopically a proofed fibre which has been dyed, or which we have attempted to dye, with logwood black, we find that the fibre presents an appearance similar to that of rope which has been drawn through some black pigment or black mud, and then dried. It is quite plain that no lustrous appearance or good "finish" can be expected from such material. Now how did the Continental hat manufacturers achieve their success, both as regards dyeing either with logwood black or with coal-tar colours, and also getting a high degree of "finish"? They attained their object by rubbing the proofing varnish on the inside of the hat bodies, in some cases first protecting the outside with a gum-varnish soluble in water but resisting the lac-varnish rubbed inside. Thus the proofing could never reach the outside. On throwing the hat bodies, thus proofed by a logical and scientific process, into the dye-bath, the gums on the outer surface are dissolved and removed, and the dye strikes into a pure, clean fibre, capable of a high degree of finish. This process, however, whilst very good for the softer hats used on the Continent, is not so satisfactory for the harder, stiffer headgear demanded in Great Britain. What was needed was a process which would allow of a through-and-through proofing and stiffening, and also of satisfactory dyeing of the stiffened and proofed felt. This was accomplished by a process patented in 1887 by Mr. F.W. Cheetham, and called the "veneering" process. The hat bodies, proofed as hard as usual, are thrown into a "bumping machine" containing hot water rendered faintly acid with sulphuric acid, and mixed with short-staple fur or wool, usually of a finer quality than that of which the hat bodies are composed. The hot acid water promotes in a high degree the felting powers of the short-staple wool or fur, and, to a lesser extent, the thinly [Pg 66] proofed ends of the fibres projecting from the surfaces of the proofed hat-forms. Thus the short-staple wool or fur felts itself on to the fibres already forming part of the hat bodies, and a new layer of pure, unproofed wool or fur is gradually wrought on to the proofed surface. The hat-forms are then taken out and washed, and can be dyed with the greatest ease and with excellent results, as will be seen from the accompanying illustration (see Fig. 15). This successful invention emphasises [Pg 67] the value of the microscope in the study of processes connected with textile fibres. I would strongly advise everyone interested in hat manufacturing or similar industries to make a collection of wool and fur fibres, and mount them on microscope slides so as to form a kind of index collection for reference.

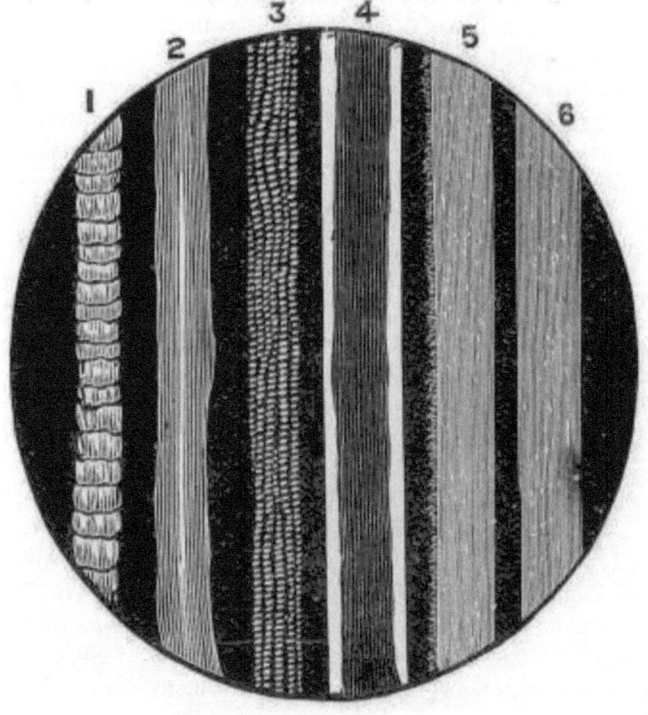

Fig. 15.

1. Natural wool fibre unproofed.

2. Wool fibre showing proof on surface, filling up the cells and rendering the same dye-proof.

3. Fur fibre from surface of veneered felt, showing dye deposited in cells and on the surface, bright and lustrous.

4. Wool fibre as in No. 2, with dye deposited on surface of proof.

5. Section of proofed and veneered body, showing unproofed surface.

6. Section of proofed body without "veneer."

[Pg 68]

LECTURE VIII

MORDANTS: THEIR NATURE AND USE

The name or word "mordant" indicates the empiricism, or our old friend "the rule of thumb," of the age in which it was first created and used. It serves as a landmark of that age, which, by the way, needed landmarks, for it was an age of something between scientific twilight and absolute darkness. *Morder* in French, derived from the Latin *mordere*, means "to bite," and formerly the users of mordants in dyeing and printing believed their action to be merely a mechanical action, that is, that they exerted a biting or corroding influence, serving to open the pores of the fabrics, and thus to give more ready ingress to the colour or dye.

Most mordants are salts, or bodies resembling salts, and hence we must commence our study of mordants by a consideration of the nature of salts. I have already told you that acids are characterised by what we term an acid reaction upon certain vegetable and artificial colours, whilst bases or basic substances in solution, especially alkalis, restore those colours, or turn them to quite another shade; the acids do the one thing, and the alkalis and soluble bases do the opposite. The strongest and most soluble bases are the alkalis—soda, potash, and ammonia. You all know, probably, that a drop of vitriol allowed to fall on a black felt hat will stain that hat red if the hat has been dyed with logwood black; and if you want to restore the black, you can do this by touching the stain with a [Pg 69] drop of strong ammonia. But the use of a black felt hat as a means of detecting acidity or alkalinity would not commend itself to an economic mind, and we find a very excellent reagent for the purpose in extract of litmus or litmus tincture, as well as in blotting paper stained therewith. The litmus is turned bright red by acids and blue by alkalis. If the acid is exactly neutralised by, that is combined with, the alkaline base to form fully neutralised salts, the litmus paper takes a purple tint. Coloured reagents such as litmus are termed indicators. A substance called phenolphthalein, a coal-tar product, is a very delicate indicator; it is more sensitive to acids than litmus is. Now there are some salts which contain a preponderance of acid in their composition, *i.e.* in which the acid has not been fully neutralised by the base; such salts are termed acid salts. Bicarbonate of soda is one of these acid salts, but so feeble is carbonic acid in its acid properties and practical evidences, that we shall see both monocarbonate or "neutral" carbonate of soda and bicarbonate or "acid" carbonate of soda show evidences of, or, as chemists say, react with alkalinity towards litmus. However, phenolphthalein, though reacting alkaline with monocarbonate of soda, indicates the acidity of the bicarbonate of soda, a thing which, as I have just said, litmus will not do. We will take two jars containing solution of monocarbonate of soda, and in the first we will put some phenolphthalein solution, and in the second, some litmus tincture. The solution in the first jar turns rose coloured, and in the second, blue, indicating in each case that the solution is alkaline. If now, however, carbonic acid be blown into the two solutions, that in the first jar, containing the phenolphthalein, becomes colourless as soon as the monocarbonate of soda is converted into bicarbonate, and this disappearance of the rose colour indicates acidity; the blue solution in the jar containing litmus, on the other hand, is not altered by blowing in carbonic acid. Furthermore, if to the [Pg 70] colourless solution containing phenolphthalein, and which is acid towards that reagent, a little reddened litmus is added, this is still turned blue, and so still indicates the presence of alkali. We have, therefore, in bicarbonate of soda a salt which behaves as an acid to phenolphthalein and as an alkali to litmus. Another extremely sensitive indicator is the coal-tar dyestuff known as "Congo red"; the colour changes produced by it are exactly the inverse of those produced in the case of litmus, that is, it gives a blue colour with acids and a red colour with alkalis.

We have now learned that acids are as the antipodes to alkalis or bases, and that the two may combine to form products which may be neutral or may have a preponderance either of acidity or of basicity — in short, they may yield neutral, acid, or basic salts. I must try to give you a yet clearer idea of these three classes of salts. Now acids in general have, as we have seen, what we may call a "chemical appetite," and each acid in particular has a "specific chemical appetite" for bases, that is, each acid is capable of combining with a definite quantity of an individual base. The terms "chemical appetite" and "specific chemical appetite" are names I have coined for your present benefit, but for which chemists would use the words "affinity" and "valency" respectively. Now some acids have a moderate specific appetite, whilst others possess a large one, and the same may be said of bases, and thus as an example we may have mono-, di-, and tri-acid salts, or mono-, di-, and tri-basic salts. In a tri-acid salt a certain voracity of the base is indicated, and in a tri-basic salt, of the acid. Again, with a base capable of absorbing and combining with its compound atom or molecule several compound atoms or molecules of an acid, we have the possibility of partial saturation, and, perhaps, of several degrees of it, and also of full saturation, which means combination to the full extent of the powers of the base in question. Also, with an acid capable [Pg 71] of, or possessing a similar large absorptive faculty for bases, we have possibilities of the formation of salts of various degrees of basicity, according to the smaller or larger degree of satisfaction given to the molecule of such acid by the addition of a base. We will now take as a simple case that of hydrochloric acid (spirits of salt), which is a monobasic acid, that is, its molecule is capable of combining with only one molecule of a monoacid base. Hydrochloric acid may be written, as its name would indicate, HCl, and an addition even of excess of such a base as caustic soda (written NaOH) would only yield what is known as common salt or chloride of sodium (NaCl), in which the metal sodium (Na) has replaced the hydrogen (H) of the hydrochloric acid. Now chloride of sodium when dissolved in water will turn litmus neither blue nor red; it is therefore neutral. Such simple, neutral, monobasic salts are mostly very stable. By "stable" we mean they possess considerable resistance to agencies, that, in the case of other salts, effect decompositions of those salts. Such other salts which are decomposed more or less readily are termed "unstable," but the terms are of course only comparative.

Now let us consider a di- or bi-basic acid. Such an one is vitriol or sulphuric acid (H_2SO_4). The hydrogen atoms are in this case an index of the basicity of the acid, and accordingly the fully saturated sodium salt is Na_2SO_4 or neutral, or better normal, sulphate of soda. In like manner the fully saturated salt of the dibasic acid, carbonic acid (H_2CO_3), is Na_2CO_3, ordinary or normal carbonate of soda. But we must observe that with these dibasic acids it is possible, by adding insufficient alkali to completely saturate them, to obtain salts in which only one hydrogen atom of the acid is replaced by the metal of the base. Thus sulphuric and carbonic acids yield $NaHSO_4$, acid sulphate or bisulphate of soda, and $NaHCO_3$, bicarbonate of soda, respectively. An example of a tribasic [Pg 72] acid is phosphoric acid, H_3PO_4, and here we may have three different classes of salts of three various degrees of basicity or base-saturation. We may have the first step of basicity due to combination with soda, NaH_2PO_4, or monosodium phosphate, the second step, Na_2HPO_4, or disodium phosphate, and the third, and final step, Na_3PO_4, or trisodium phosphate. Now let us turn to the varying degrees of acidity, or rather the proportions of acid radicals in salts, due to the varying appetites or combining powers of bases. Sodium only forms simple monoacid salts, as sodium chloride (NaCl), sodium sulphate (Na_2SO_4); calcium forms diacid salts, *e.g.* calcium chloride ($CaCl_2$); and aluminium and iron, triacid salts, for example, aluminium sulphate [$Al_2(SO_4)_3$] and iron (ferric) sulphate [$Fe_2(SO_4)_3$]. Now in these triacid salts we can remove some of the acid groups and substitute the elements of water, OH, or hydroxyl, as it is called, for them. Such salts, then, only partly saturated with acid, are termed basic salts. Thus we have $Al_2(OH)_2(SO_4)_2$, $Al_2(OH)_4SO_4$, as well as $Al_2(SO_4)_3$, and we can get these basic salts by treating the normal sulphate [$Al_2(SO_4)_3$] with sufficient caustic soda to remove the necessary quantities of sulphuric acid. Now it is a curious thing that of these aluminium sulphates the fully saturated one, $Al_2(SO_4)_3$, is the most stable, for even on long boiling of its solution in water it suffers no change, but the more basic is the sulphate the less stable it becomes, and so the more easily it decomposes on heating or boiling its solution, giving a deposit or precipitate of a still more basic sulphate, or of hydrated alumina itself, $Al_2(OH)_6$, until we arrive at the salt $Al_2(SO_4)_2(OH)_2$, which is quite unstable on boiling; $Al_2(SO_4)(OH)_4$ would be more unstable still. This behaviour may be easily shown experimentally. We will dissolve some "cake alum" or normal sulphate of alumina, $Al_2(SO_4)_3$, in water, and boil some of the solution. No deposit or precipitate is produced; the salt is stable. To another portion of the solution we will add some caustic soda, NaOH, [Pg 73] in order to rob the normal sulphate of alumina of some of its sulphuric acid. This makes the sulphate of alumina basic, and the more basic, the more caustic soda is added, the sodium (Na) of the caustic soda combining with the SO_4 of the sulphate of alumina to form sulphate of soda (Na_2SO_4), whilst the hydroxyl (OH) of the caustic soda takes the position previously occupied by the SO_4. But this increase of basicity also means decrease of stability, for on boiling the solution, which now contains a basic sulphate of alumina, a precipitate is formed, a result which also follows if more caustic soda is added, production of still more basic salts or of hydrated alumina, $Al_2(OH)_6$, taking place in either case.

Mordanting or Fixing Acid (Phenolic) Colours.—But what has all this to do with mordanting? is possibly now the inquiry. So much as this, that only such unstable salts as I have just described, which decompose and yield precipitates by the action on them of alkalis, heat, the textile fibres themselves, or other agencies, are suitable to act as true mordants. Hence, generally, the sources or root substances of the best and most efficient mordants are the metals of high specific appetite or valency. I think we have now got a clue to the principle of mordants and also to the importance of a sound chemical knowledge in dealing most effectively with them, and I may tell you that the man who did most to elucidate the theory of mordanting is not a practical man in the general sense of the term, but a man of the highest scientific attainments and standing, namely, Professor Liechti, who, with his colleague Professor Suida, did probably more than any other man to clear up much that heretofore was cloudy in this region. We have seen that with aluminium sulphate, basic salts are precipitated, *i.e.* salts with such a predominance of appetite for acids, or such *quasi*-acids as phenolic substances, that if such bodies were present they would combine with the basic parts of those precipitated salts as soon as the latter were formed, and [Pg 74] all would be precipitated together as one complex compound. Just such peculiar *quasi*-acid, or phenolic substances are Alizarin, and most of the natural adjective dyestuffs, the colouring principles of logwood, cochineal, Persian berries, etc. Hence these substances will be combined and carried down with such precipitated basic salts. The complex compounds thus produced are coloured substances known as lakes. For example, if I take a solution containing basic sulphate of alumina, prepared as I have already described, and add to some Alizarin, and then heat the mixture, I shall get a red lake of Alizarin and alumina precipitated. If I had taken sulphate of iron instead of sulphate of alumina, and proceeded in a similar manner, and added Alizarin, I should have obtained a dark purple lake. Now if you imagine these reactions going on in a single fibre of a textile material, you have grasped the theory and purpose of mordanting. The textile fabric is drawn through the alumina solution to fill the pores and tubes of the fabric; it is then passed through a weak alkaline bath to basify or render basic the aluminium salt in the pores; and then it is finally carried into the dye-bath and heated there, in order to precipitate the colour lake in the fibre. The method usually employed to mordant woollen fabrics consists in boiling them with weak solutions of the metallic salts used as mordants, often with the addition of acid salts, cream of tartar, and the like. A partial decomposition of the metallic salts ensues, and it is induced by several conditions: (1) The dilution of the liquid; (2) the heating of the solution; (3) the presence of the fibre, which itself tends to cause the breaking up of the metallic salts into less soluble basic ones. Thus it is not really necessary to use basic aluminium sulphate for mordanting wool, since the latter itself decomposes the normal or neutral sulphate of alumina on heating, an insoluble basic sulphate being precipitated in the fibres of the wool. (4) The presence of other added substances, as cream of tartar, etc. The best alumina mordant is probably the acetate of alumina [Pg 75] ("red liquor"), and the best iron mordant, probably also the acetate ("iron liquor") (see preceding lecture), because the acetic acid is so harmless to the fibre, and is easily driven off on steaming, etc. A further reason is that from the solution of acetate of iron or alumina, basic acetates are very easily precipitated on heating, and are thus readily deposited in the fibre.

Mordanting and Fixing Basic Colours.—Now let us ask ourselves a very important question. Suppose we have a colour or dyestuff, such as Magenta, which is of a basic character, and not of an acid or phenolic character like the colours Alizarin, Hæmatein (logwood), or carminic acid (cochineal), and we wish to fix this basic dyestuff on the tissue. Can we then use "red liquor" (acetate of alumina), acetate of iron, copperas, etc.? The answer is, No; for such a process would be like trying to combine base with base, instead of base with acid, in order to form a salt. Combination, and so precipitation, would not take place; no lake would be formed. We must seek for an acid or acid body to use as mordant for our basic colour, and an acid or acid body that will form an insoluble precipitate or colour-lake with the dyestuff. An acid much used, and very valuable for this purpose, is tannic acid. The tannate of rosaniline (colour principle of Magenta) is a tolerably insoluble lake, which can be precipitated by Magenta from a solution of tannate of soda, the Magenta being capable of displacing the soda. But tannic acid, alone, does not form very fast lakes with Magenta and the other basic dyestuffs, and so a means of rendering these lakes more insoluble is needed. It is found that tannic acid and tartar emetic (a tartrate of antimony and potash) yield a very insoluble compound, a tannate of antimony. Perchloride of tin, in a similar manner, yields insoluble tannate of tin with tannic acid. These insoluble compounds, however, have sufficient acid-affinity left in the combined tannic acid to unite also with the basic aniline colours, forming very fast or insoluble colour lakes. This [Pg 76] principle is extensively used in practice to fix basic aniline colours, especially on cotton. We should first soak the piece of cotton in a solution of tannic acid, and then pass it into a solution, say, of tartar emetic, when the tannic acid will be firmly fixed, as tannate of antimony, on the cotton. We then dip the mordanted piece of cotton into the colour bath, containing, for instance, Magenta, and it is dyed a fine red, composed of a tannate of antimony and Magenta. You now see, no doubt, the necessity of sharply discriminating between two classes of colouring matters, which we may term *colour acids* and *colour bases* respectively. There are but few acids that act like tannic acid in fixing basic aniline dyestuffs, but oleic acid and other fatty acids are of the number. A curious question might now be asked, namely: "Could the acid colour Alizarin, if fixed on cotton cloth, combine with a basic aniline colour, *e.g.* Aniline Violet, and act as a mordant for it, thus fixing it?" The answer is, "Certainly"; and thus an Alizarin Purple would be produced, whilst if Magenta were used in place of Aniline Violet, an Alizarin Red of a crimson tone would result.

Chrome Mordanting of Wool and Fur.—In studying this subject I would recommend a careful perusal of the chapter on "Mordants" in J.J. Hummel's book, entitled *The Dyeing of Textile Fabrics*, and pages 337 to 340 of Bowman's work on *The Wool-Fibre*.

In the treatment of wool or fur with bichrome (potassium bichromate) we start with an acid salt, a bichromate ($K_2Cr_2O_7$) and a strong oxidising agent, and we finish with a basic substance, namely, oxide of chromium, in the fibres of the wool or fur. If we desire to utilise the whole of the chromic acid in our mordanting liquor, we must add to it some sulphuric acid to set free the chromic acid from the potassium with which it is combined. Bichromate of potash with sulphuric acid gives sulphate of potash and chromic acid. The question [Pg 77] of the proper exhaustion of bichromate baths is an important economic one. Now we must remember that this chromic acid (CrO_3) oxidises our wool or fur, and must oxidise it before it can of itself act as a mordant by being reduced in the process to hydrated chromic oxide, $Cr_2O_3 + 3\ H_2O$. [$2\ CrO_3$ (chromic acid) = Cr_2O_3 (chromic oxide) + O_3 (oxygen).] It is this hydrated chromic oxide in the fibre that yields with the Hæmatein of the logwood your logwood black dye. Mr. Jarmain finds that it is not safe to use more than 3 per cent. (of the weight of the wool) of bichromate; if 4 per cent. be used, the colour becomes impaired, whilst if 12 per cent. be employed, the wool cannot be dyed at all with logwood, the phenomenon known as "over-chroming" being the result of such excessive treatment. I think there is no doubt, as Professor Hummel says, that the colouring matter is oxidised and destroyed in such over-chroming, but I also think that there can be no doubt that the wool itself is also greatly injured and incapacitated for taking up colour. Now the use of certain coal-tar black dyes in place of logwood obviates this use of bichrome, and thus the heavy stress on the fibre in mordanting with it. It also effects economy in avoiding the use of bichrome, as well as of copper salts; but even thus, of course, other problems have to be solved before it can be finally decided which is best.

[Pg 78]

LECTURE IX

DYESTUFFS AND COLOURS

Classification.—In classifying the different dyestuffs and colouring matters it is, of course, necessary to consider first the properties of those colouring matters generally, and secondly the particular reason for making such classification. The scientific chemist, for example, would classify them according to theoretical considerations, as members of certain typical groups; the representative of medical science or hygiene would naturally classify them as poisonous and non-poisonous bodies; whilst the dyer will as naturally seek to arrange them according to their behaviour when applied to textile fabrics. But this behaviour on applying to textile fibres, if varied in character according to the chemical nature of the colouring matter, as well as the chemical and physical nature of the fabric—and it is so varied—will make such classification, if it is to be thoroughgoing, not a very simple matter. I may tell you that it is not a simple matter, and, moreover, the best classification and arrangement is that one which depends both on the action of the dyes on the fibres, and also on the intrinsic chemical character of the dyestuffs themselves. Since the higher branches of organic chemistry are involved in the consideration of the structure and dispositions, and consequently more or less of the properties of these dyes, you will readily comprehend that the thorough appreciation and use of that highest and best method of classification, particularly in the [Pg 79] case of the coal-tar dyes, will be, more or less, a sealed book except to the student of organic chemistry. But it may be asked, "How does that highest and best method of classifying the dyestuffs affect the users, the dyers, in their processes?" In reply, I would say, "I believe that the dyer who so understands the chemical principles involved in the processes he carries out, and in the best methods of classifying the dyes as chemical substances, so as to be able to act independently of the prescriptions and recipes given him by the dye manufacturers, and so be master of his own position, will, *ceteris paribus*, be the most economical and successful dyer." Many manufacturers of dyestuffs have said the very same thing to me, but, independently of this, I know it, and can prove it with the greatest ease. Let me now, by means of an experiment or two, prove to you that at least some classification is necessary to begin with. So different and varied are the substances used as colouring matters by the dyer, both as regards their chemical and physical properties, that they even act differently towards the same fibre. I will take four pieces of cotton fabric; three of them are simple white cotton, whilst the fourth cotton piece has had certain metallic salts mixed with thickening substances like gum, printed on it in the form of a pattern, which at present cannot readily be discerned. We will now observe and note the different action on these pieces of cotton—(i.) of a Turmeric bath, (ii.) a Magenta bath, and (iii.) a madder or Alizarin bath. The Turmeric dyes the cotton a fast yellow, the Magenta only stains the cotton crimson, and on washing with water alone, almost every trace of colour is removed again; the madder, however, stains the cotton with no presentable shade of colour at all, produces a brownish-yellow stain, removed at once by a wash in water. But let us take the printed piece of cotton and dye that in the Alizarin bath, and then we shall discover the conditions for producing colours with such a dyestuff as madder or Alizarin. [Pg 80] Different coloured stripes are produced, and the colours are conditioned by the kind of metallic salts used. Evidently the way in which, the turmeric dyes the cotton is different from that in which the madder dyes it. The first is a yellow dyestuff, but it would be hard to assign any one shade or tint to Alizarin as a dyestuff. In fact Alizarin (the principle of madder) is of itself not a dye, but it forms with each of several metals a differently coloured compound; and thus the metallic salt in the fabric is actually converted into a coloured compound, and the fabric is dyed or printed. The case is just the same with logwood black dyeing: without the presence of iron ("copperas," etc.), sulphate of copper ("bluestone"), or bichrome, you would get no black at all. We will now try similar experiments with woollen fabrics, taking three simple pieces of flannel, and also two pieces, the one having been first treated with a hot solution of alum and cream of tartar, and the other with copperas or sulphate of iron solution, and then washed. Turmeric dyes the first yellow, like it did the cotton. Magenta, however, permanently dyes the woollen as it did not the cotton. Alizarin only stains the untreated woollen, whilst the piece treated with alumina is dyed red, and that with iron, purple. If, however, the pieces treated with iron and alumina had been dyed in the Magenta solution, only one colour would have been the result, and that a Magenta-red in each case. Here we have, as proved by our experiments, two distinct classes of colouring matters. The one class comprises those which are of themselves the actual colour. The colour is fully developed in them, and to dye a fabric they only require fixing in their unchanged state upon that fabric. Such dyes are termed *monogenetic*, because they can only generate or yield different shades of but one colour. Indigo is such a dye, and so are Magenta, Aniline Black, Aniline Violet, picric acid, Ultramarine Blue, and so on. Ultramarine is not, it is true, confined to blue; you can get Ultramarine Green, and even rose-coloured Ultramarine; but [Pg 81] still, in the hands of the dyer, each shade remains as it came from the colour-maker, and so Ultramarine is a monogenetic colour. Monogenetic means capable of generating one. Turning to the other class, which comprises, as we have shown, Alizarin, and, besides, the colouring principle of logwood (Hæmatein), Gallein, and Cochineal, etc., we have bodies usually possessed of some colour, it is true, but such colour is of no consequence, and, indeed, is of no use to dyers. These bodies require a special treatment to bring out or develop the colours, for there may be several that each is capable of yielding. We may consider them as colour-giving principles, and so we term them *polygenetic* colours. Polygenetic means capable of generating several or many. In the various colours and dyes we have all phases, and the monogenetic shades almost imperceptibly into the polygenetic. The mode of application of the two classes of colours is, of course, in each case quite essentially different, for in the case of the monogenetic class the idea is mainly either to dye at once and directly upon, the unprepared fibre, or having subjected the fabric to a previous preparation with a metallic or other solution, to fix directly the one colour on that fabric, on which, without such preparation, it would be loose. In the case of the polygenetic class, the idea is necessarily twofold. The dyeing materials are not colours, only colour generators. Hence in all cases the fabric must be prepared with the twofold purpose—first, of using a metallic or other agent, capable of yielding, with the dye material, the desired colour; and secondly, of yielding it on the fibre in an insoluble and permanent form. Now, though I have gone so far into this mode of classification, because it does afford some information and light, yet I can go no farther without getting into a territory that presupposes a knowledge and acquaintance with the chemical structure of the colouring matters as organic substances, which would be, at present, beyond us. I shall now turn to another mode of [Pg 82] classification, which, if not so far-reaching as the other, is at least an exceedingly useful one. The two methods may be combined to a considerable extent. By the latter plan the colours may be conveniently divided into three groups: I., substantive colours; II., adjective colours; III., mineral and pigment colours.

Substantive Dyestuffs. — The substantive colours fix themselves readily and directly on animal fibres and substances, but only a few amongst them will dye vegetable fibres like cotton and linen directly. Almost all substantive colours may, however, be fixed on cotton and linen by first preparing or mordanting those vegetable fibres. Silk, wool, fur, etc., act like fibre and mordant together, for they absorb and fix the substantive colours firmly. In our experiments we saw that turmeric is one of the few substantive colours fixing itself on both cotton and wool, without any aid from a mordant or fixing agent. Magenta was also a substantive colour, but Alizarin was certainly not one of this class.

Adjective Dyestuffs. — Some of these substances are definitely coloured bodies, but in some of them the colour is of no consequence or value, and is quite different and distinct from the colour eventually formed on the fibre, which colour only appears in conjunction with a special mordant; but, again, some of them are not coloured, and would not colour the fibre directly at all, only in conjunction with some mordant. All the polygenetic colours are, of course, comprised in this class, for example Alizarin and logwood (Hæmatein), whilst such monogenetic colours as annatto and turmeric are substantive, for they will fix themselves without a mordant on cotton and wool. The adjective colours can be conveniently subdivided into — (*a*) those existing in nature, as logwood (Hæmatein) and Cochineal; (*b*) those artificially formed from coal-tar products, as Alizarin (madder), Gallein, etc.

Mineral and Pigment Dyestuffs. — These colours are insoluble [Pg 83] in water and alcohol. They are either fixed on the fibre by mechanical means or by precipitation. For example, you use blacklead or plumbago to colour or darken your hats, and you work on this pigment colour by mechanical means. I will show you by experiment how to fix a coloured insoluble pigment in the fibre. I take a solution of acetate of lead (sugar of lead), and to it I add some solution of bichrome (potassium bichromate). Acetate of lead (soluble in water) with bichromate of potash (also soluble in water) yields, on mixing the two, acetate of potash (soluble in water), and chromate of lead, or chrome yellow (insoluble in water), and which is consequently precipitated or deposited. Now suppose I boil some of that chrome-yellow precipitate with lime-water, I convert that chrome yellow into chrome orange. This, you see, takes place without any reference to textile fibres. I will now work a piece of cotton in a lead solution, so that the little tubes of the cotton fibre shall be filled with it just as the larger glass tube or vessel was filled in the first experiment. I next squeeze and wash the piece, so as to remove extraneous solution of lead, just as if I had filled my glass tube by roughly dipping it bodily into the lead solution, and then washed and cleansed the outside of that tube. Then I place the fabric in a warm solution of bichromate of potash (bichrome), when it becomes dyed a chrome yellow, for just as chromate of lead is precipitated in the glass tube, so it is now precipitated in the little tubes of the cotton fibre (see Lecture I.). Let us see if we can now change our chrome yellow to chrome orange, just as we did in the glass vessel by boiling in lime-water. I place the yellow fabric in boiling lime-water, when it is coloured or dyed orange. In each little tubular cotton fibre the same change goes on as went on in the glass vessel, and as the tube or glass vessel looks orange, so does the fabric, because the cotton fibres or tubes are filled with the orange chromium compound. You see this is quite a different process of pigment colouring [Pg 84] from that of rubbing or working a colour mechanically on to the fibre.

Let us now turn to the substantive colours (Group I.), and see if we can further sub-divide this large group for the sake of convenience. We can divide the group into two—(*a*) such colours as exist ready formed in nature, and chiefly occur in plants, of which the following are the most important: indigo, archil or orchil, safflower, turmeric, and annatto; (*b*) the very large sub-group of the artificial or coal-tar colours. We will briefly consider now the dyestuffs mentioned in Group (*a*).

Natural Substantive Colours.—Indigo, one of the most valuable dyes, is the product of a large number of plants, the most important being different species of *indigofera*, which belong to the pea family. None of the plants (of which *indigofera tinctoria* is the chief) contain the colouring matter in the free state, ready-made, so to say, but only as a peculiar colourless compound called *indican*, first discovered by Edward Schunck. When this body is treated with dilute mineral acids it splits up into Indigo Blue and a kind of sugar. But so easily is this change brought about that if the leaf of the plant be only bruised, the decomposition ensues, and a blue mark is produced through separation of the Indigo Blue. The possibility of dyeing with Indigo so readily and easily is due to the fact that Indigo Blue absorbs hydrogen from bodies that will yield it, and becomes, as we say, reduced to a body without colour, called Indigo White, a body richer in hydrogen than Indigo Blue, and a body that is soluble. If this white body (Indigo White) be exposed to the air, the oxygen of the air undoes what the hydrogen did, and oxidises that Indigo White to insoluble Indigo Blue. Textile fabrics dipped in such reduced indigo solutions, and afterwards exposed to the air, become blue through deposit in the fibres of the insoluble Indigo Blue, and are so dyed. This is called the indigo-vat method. We can reduce this indigo so as to prepare the indigo-vat [Pg 85] by simply mixing Indigo Blue, copperas (ferrous sulphate) solution, and milk of lime in a closely-stoppered bottle with water, and letting the mixture stand. The clear liquor only is used. A piece of cotton dipped in it, and exposed to the air, quickly turns blue by absorbing oxygen, and is thus dyed. The best proportions for the indigo-vat are, for cloth dyeing, 4000 parts of water, 40 of indigo, 60 to 80 of copperas crystals, and 50 to 100 of dry slaked lime. The usual plan is to put in the water first, then add the indigo and copperas, which should be dissolved first, and finally to add the milk of lime, stirring all the time. Artificial indigo has been made from coal-tar products. The raw material is a coal-tar naphtha called toluene or toluol, which is also the raw material for saccharin, a sweetening agent made from coal-tar. This artificial indigo is proving a formidable rival to the natural product.

Orchil paste, orchil extract, and cudbear are obtained by exposing the plants (species of lichens) containing the colouring principle, called *Orcin*, itself a colourless substance, to the joint action of ammonia and air, when the oxygen of the air changes that orcin by oxidising it into *Orcèin*, which is the true red colouring matter contained in the preparations named. The lichens thus treated acquire gradually a deep purple colour, and form the products called "cudbear." This dye works best in a neutral bath, but it will do what not many dyes will, namely, dye in either a slightly alkaline or slightly acid bath as well. Orchil is not applicable in cotton dyeing. Being a substantive colour no mordants are needed in dyeing silk and wool with it. The colour produced on wool and silk is a bright magenta-red with bluish shade.

Litmus is also obtained from the same lichens as yield orchil. It is not used in dyeing, and is a violet-blue colouring matter when neither acid nor alkaline, but neutral as it is termed. It turns red with only a trace of acid, and blue with the least trace of alkali, and so forms a very delicate reagent when pieces [Pg 86] of paper are soaked with it, and dipped into the liquids to be tested.

Safflower: This vegetable dyeing material, for producing pink colours on cotton without the aid of a mordant, consists of the petals of the flower of *carthamus tinctorius*. It contains a principle termed "Carthamin" or "carthamic acid," which can be separated by exhausting safflower with cold acidulated water (sulphuric acid) to dissolve out a yellow colouring matter which is useless. The residue after washing free from acid is treated with a dilute solution of soda crystals, and the liquid is then precipitated by an acid. A red precipitate is obtained, which fixes itself directly on cotton thread immersed in the liquid, and dyes it a delicate rose pink, which is, unfortunately, very fugitive. Silk can be dyed like cotton. The colour is not fast against light.

Turmeric is the root portion of a plant called *curcuma tinctoria*, that grows in Southern Asia. The principle forming the colouring matter is "Curcumin." It is insoluble in cold water, not much soluble in hot, but easily soluble in alcohol. From the latter solution it separates in brilliant yellow crystals. Although the colour it yields is very fugitive, the wool and silk dyers still use it for producing especially olives, browns, and similar compound shades. It produces on cotton and wool a bright yellow colour without the aid of any mordant. To show you how easily dyeing with turmeric is effected, I will warm some powdered turmeric root in a flask with alcohol, and add the extract to a vessel of water warmed to about 140° F. (60° C.), and then dip a piece of cotton in and stir it about, when it will soon be permanently dyed a fine bright yellow. A piece of wool similarly worked in the bath is also dyed. However, the unfortunate circumstance is that this colour is fast neither to light nor alkalis. Contact with soap and water, even, turns the yellow-dyed cotton, reddish-brown.

Annatto is a colouring principle obtained from the pulpy [Pg 87] matter enclosing the seeds of the fruit of a tree, the *Bixa orellana*, growing in Central and Southern America. The red or orange colour it yields is fugitive, and so its use is limited, being chiefly confined to silk dyeing. The yellow compound it contains is called "Orellin," and it also contains an orange compound called "Bixin," which is insoluble in water, but readily soluble in alkalis and in alcohol with a deep yellow colour. To dye cotton with it, a solution is made of the colour in a boiling solution of carbonate of soda. The cotton is worked in the diluted alkaline solution whilst hot. By passing the dyed cotton through water acidulated with a little vitriol or alum, a redder tint is assumed. For wool and silk, pale shades are dyed at 106° F. (50° C.) with the addition of soap to the bath, dark shades at 200° to 212° F. (80° to 100° C.).

[Pg 88]

LECTURE X

DYESTUFFS AND COLOURS — *Continued*

Artificial Substantive Dyestuffs. — You may remember that in the last lecture we divided the colouring matters as follows: I. Substantive colours, fixing themselves directly on animal fibres without a mordant, only a few of them doing this, however, on vegetable fibres, like cotton. We sub-divided them further as — (*a*) those occurring in nature, and (*b*) those prepared artificially, and chiefly, but not entirely, the coal-tar colouring matters. II. Adjective colours, fixing themselves only in conjunction with a mordant or mordants on animal or vegetable fibres, and including all the polygenetic colours. III. Mineral or pigment colours. I described experiments to illustrate what we mean by monogenetic and polygenetic colours, and indicating that the monogenetic colours are mainly included in the group of substantive colours, whilst the polygenetic colours are mainly included in the adjective colours. But I described also an illustration of Group III., the mineral or pigment colours, by which we may argue that chromate of lead is a polygenetic mineral colour, for, according to the treatment, we were able to obtain either chrome yellow (neutral lead chromate) or chrome orange (basic lead chromate). I also said there was a kind of borderland whichever mode of classification be adopted. Thus, for example, there are colours that are fixed on the fibre either directly like indigo, and so are substantive, or they may be, and generally are, [Pg 89] applied with a mordant like the adjective and polygenetic colours; examples of these are Cœrulein, Alizarin Blue, and a few more. We have now before us a vast territory, namely, that of the *b* group of substantive colours, or, the largest proportion, indeed almost all of those prepared from coal-tar sources; Alizarin, also prepared from coal-tar, belongs to the adjective colours. With regard to the source of these coal-tar colours, the word "coal-tar," I was going to say, speaks volumes, for the destructive and dry distillation of coal in gas retorts at the highest temperatures to yield illuminating gas, also yields us tar. But, coal distilled at lower temperatures, as well as shale, as in Scotland, will yield tar, but tar of another kind, from which colour-generating substances cannot be obtained practically, but instead, paraffin oil and paraffin wax for making candles, etc. Coal-tar contains a very large number of different substances, but only a few of them can be extracted profitably for colour-making. All the useful sources of colours and dyes from coal-tar are simply compounds of carbon and hydrogen—hydrocarbons, as they are called, with the exception of one, namely, phenol, or carbolic acid. I am not speaking here of those coal-tar constituents useful for making dyes, but of those actually extracted from coal-tar for that purpose, *i.e.* extracted to profit. For example, aniline is contained in coal-tar, but if we depended on the aniline contained ready made in coal-tar for our aniline dyes, the prices of these dyes would place them beyond our reach, would place them amongst diamonds and precious stones in rarity and cost, so difficult is it to extract the small quantity of aniline from coal-tar. The valuable constituents actually extracted are then these: benzene, toluene, xylene, naphthalene, anthracene, and phenol or carbolic acid. One ton of Lancashire coal, when distilled in gas retorts, yields about 12 gallons of coal-tar. Let us now learn what those 12 gallons of tar will give us in the shape of hydrocarbons and carbolic acid, mentioned as extracted profitably [Pg 90] from tar. This is shown very clearly in the following table (Table A).

The 12 gallons of tar yield 1-1/10 lb. of benzene, 9/10 lb. of toluene, 1½ lb. of carbolic acid, between 1/10 and 2/10 lb. of xylene, 6½ lb. of naphthalene, and ½ lb. of anthracene, whilst the quantity of pitch left behind is 69½ lb. But our table shows us more; it indicates to us what the steps are from each raw material to each colouring matter, as well as showing us each colouring matter. We see here that our benzene yields us an equal weight of aniline, and the toluene (9/10 lb.) about 3/4 lb. of toluidine, the mixture giving, on oxidation, between ½ and 3/4 lb of Magenta. From carbolic acid are obtained both Aurin and picric acid, and here is the actual quantity of Aurin obtainable (1-1/4 lb.). From naphthalene, either naphthylamine (a body like aniline) or naphthol (resembling phenol) may be prepared. The amounts obtainable you see in the table. There are two varieties of naphthol, called alpha- and beta-naphthol, but only one phenol, namely, carbolic acid. Naphthol Yellow is of course a naphthol colour, whilst Vermilline Scarlet is a dye containing both naphthylamine and naphthol. You see the quantities of these dyes, namely 7 lb. of Scarlet and 9½ lb. of the Naphthol Yellow. The amount of pure anthracene obtained is ½ lb. This pure anthracene exhibits the phenomenon of fluorescence, that is, it not only looks white, but when the light falls on it, it seems to reflect a delicate violet or blue light. Our table shows us that from the 12 gallons of tar from 1 ton of coal we may gain 2-1/4 lb. of 20 per cent. Alizarin paste. Chemically pure Alizarin crystallises in bright-red needles; it is the colouring principle of madder, and also of Alizarin paste. But the most wonderful thing about substantive coal-tar colours is their immense tinctorial power, *i.e.* the very little quantity of each required compared with the immense superficies of cloth it will dye to a full shade.

[Pg 91]

TABLE A. [1]

Twelve Gallons of Gas-Tar (average of Manchester and Salford Tar) yield:—

Toluene.	Phenol.	Solvent Naphtha for India rubber, containing the three Xylenes.	Heavy Naphtha.	Naphthalene.	Creosote.	Heavy Oil
0·90 lb.= 0·77 lb. of Toluidine.	1·5 lb. = 1·2 lb. of Aurin.	2·44 lb., yielding 0·12 lb. of Xylene = 0·07 lb. of Xylidine.	2·40 lb.	6·30 lb. = 5·25 lb. of α-Naphthylamine= 7·11 lb. of Vermilline Scarlet RRR; or 4·75 lb. of α- or β- Naphthol = 9·50 lb. of Naphthol Yellow.	17 lb.	14 lb.

[1] This table was compiled by Mr. Ivan Levinstein, of Manchester.

[Pg 92]

The next table (see Table B) shows you the dyeing power of the colouring matters derived from 1 ton of Lancashire coal, which will astonish any thoughtful mind, for the Magenta will dye 500 yards of flannel, the Aurin 120 yards, the Vermilline Scarlet 2560 yards, and the Alizarin 255 yards (Turkey-red cotton cloth).

The next table (Table C) shows the latent dyeing power resident, so to speak, in 1 lb. of coal.

By a very simple experiment a little of a very fine violet dye can be made from mere traces of the materials. One of the raw materials for preparing this violet dye is a substance with a long name, which itself was prepared from aniline. This substance is tetramethyldiamidobenzophenone, and a little bit of it is placed in a small glass test-tube, just moistened with a couple of drops of another aniline derivative called dimethylaniline, and then two drops of a fuming liquid, trichloride of phosphorus, added. On simply warming this mixture, the violet dyestuff is produced in about a minute. Two drops of the mixture will colour a large cylinder of water a beautiful violet. The remainder (perhaps two drops more) will dye a skein of silk a bright full shade of violet. Here, then, is a magnificent example of enormous tinctorial power. I must now draw the rein, or I shall simply transport you through a perfect wonderland of magic, bright colours and apparent chemical conjuring, without, however, an adequate return of solid instruction that you can carry usefully with you into every-day life and practice.

TABLE B. [1]

Dyeing Powers of Colours from 1 Ton of Lancashire Coal.

0·623 lb. of Magenta will dye 500 yards of flannel, 27 inches wide, a full shade.	1·23 lb. of Methyl Violet will dye 1000 yards of flannel, 27 inches wide, a full violet.	9.5 lb. of Naphthol Yellow will dye 3800 yards of flannel, 27 inches wide, a full yellow.	7·11 lb. of Vermilline will will dye 2560 yards of flannel, 27 inches wide, a full scarlet.	1·2 lb. of Aurin will dye 120 yards of flannel, 27 inches wide, a full orange.	2·25 lb. of Alizarin (20%) will dye 255 yards of Printers' cloth a full Turkey red.

TABLE C.

Dyeing Powers of Colours from 1 Lb. of Lancashire Coal.

Magenta or	Methyl Violet.	Naphthol Yellow or	Vermilline Scarlet.	Aurin (Orange).	Alizarin (Turkey Red).
8 × 27 inches of flannel.	24 × 27 inches of flannel.	61 × 27 inches of flannel.	41 × 27 inches of flannel.	1·93 × 27 inches of flannel.	4 × 27 inches of Printers' cloth.

[1] These tables were compiled by Mr. Ivan Levinstein, of Manchester.

Before we go another step, I must ask and answer, therefore, a few questions. Can we not get some little insight into the structure and general mode of developing the leading coal-tar colours which serve as types of whole series? I will try what can be done with the little knowledge of chemistry we have so far accumulated. In our earlier lectures we have learnt that water is a compound of hydrogen and oxygen, and in its compound atom or molecule we have two atoms of hydrogen [Pg 93] [Pg 94] combined with one of oxygen, symbolised as H_2O. We also learnt that ammonia, or spirits of hartshorn, is a compound of hydrogen with nitrogen, three atoms of hydrogen being combined with one of nitrogen, thus, NH_3. An example of a hydrocarbon or compound of carbon and hydrogen, is marsh gas (methane) or firedamp, CH_4. Nitric acid, or *aqua fortis*, is a compound of nitrogen, oxygen, and hydrogen, one atom of the first to three of the second and one of the third—NO_3H. But this nitric acid question forces me on to a further statement, namely, we have in this formula or symbol, NO_3H, a twofold idea—first, that of the compound as a whole, an acid; and secondly, that it is formed from a substance without acid properties by the addition of water, H_2O, or, if we like, HOH. This substance contains the root or radical of the nitric acid, and is NO_2, which has the power of replacing one of the hydrogen atoms, or H, of water, and so we get, instead of HOH, NO_2OH, which is nitric acid. This is chemical replacement, and on such replacement depends our powers of building up not only colours, but many other useful and ornamental chemical structures. You have all heard the old-fashioned statement that "Nature abhors a vacuum." We had a very practical example of this when in our first lecture on water I brought an electric spark in contact with a mixture of free oxygen and hydrogen in a glass bulb. These gases at once united, three volumes of them condensing to two volumes, and these again to a minute particle of liquid water. A vacuum was left in that delicate glass bulb whilst the pressure of the atmosphere was crushing with a force of 15 lb. on the square inch on the outside of the bulb, and thus a violent crash was the result of Nature's abhorrence. There is such a kind of thing, though, and of a more subtle sort, which we might term a chemical vacuum, and it is the result of what we call chemical valency, which again might be defined as the specific chemical appetite of each substance.

[Pg 95]

Let us now take the case of the production of an aniline colour, and let us try to discover what aniline is, and how formed. I pointed to benzene or benzol in the table as a hydrocarbon, C_6H_6, which forms a principal colour-producing constituent of coal-tar. If you desire to produce chemical appetite in benzene, you must rob it of some of its hydrogen. Thus C_6H_5 is a group that would exist only for a moment, since it has a great appetite for H, and we may say this appetite would go the length of at once absorbing either one atom of H (hydrogen) or of some similar substance or group having a similar appetite. Suppose, now, I place some benzene, C_6H_6, in a flask, and add some nitric acid, which, as we said, is NO_2OH. On warming the mixture we may say a tendency springs up in that OH of the nitric acid to effect union with an H of the C_6H_6 (benzene) to form HOH (water), when an appetite is at once left to the remainder, C_6H_5—on the one hand, and the NO_2—on the other, satisfied by immediate union of these residues to form a substance $C_6H_5NO_2$, nitro-benzene or "essence of mirbane," smelling like bitter almonds. This is the first step in the formation of aniline.

I think I have told you that if we treat zinc scraps with water and vitriol, or water with potassium, we can rob that water of its oxygen and set free the hydrogen. It is, however, a singular fact that if we liberate a quantity of fresh hydrogen amongst our nitrobenzene $C_6H_5NO_2$, that hydrogen tends to combine, or evinces an ungovernable appetite for the O_2 of that NO_2 group, the tendency being again to form water H_2O. This, of course, leaves the residual C_6H_5N: group with an appetite, and only the excess of hydrogen present to satisfy it. Accordingly hydrogen is taken up, and we get $C_6H_5NH_2$ formed, which is aniline. I told you that ammonia is NH_3, and now in aniline we find an ammonia derivative, one atom of hydrogen (H) being replaced by the group C_6H_5. I will now describe the method of preparation of a small quantity of aniline, in order to [Pg 96] illustrate what I have tried to explain in theory. Benzene from coal-tar is warmed with nitric acid in a flask. A strong action sets in, and on adding water, the nitrobenzene settles down as a heavy oil, and the acid water can be decanted off. After washing by decantation with water once or twice, and shaking with some powdered marble to neutralise excess of acid, the nitrobenzene is brought into contact with fresh hydrogen gas by placing amongst it, instead of zinc, some tin, and instead of vitriol, some hydrochloric acid (spirits of salt). To show you that aniline is formed, I will now produce a violet colour with it, which only aniline will give. This violet colour is produced by adding a very small quantity of the aniline, together with some bleaching powder, to a mixture of chalk and water, the chalk being added for the purpose of destroying acidity. This aniline, $C_6H_5NH_2$, is a base, and forms the foundation of all the so-called basic aniline colours. If I have made myself clear so far, I shall be contented. It only remains to be said that for making Magenta, pure aniline will not do, what is used being a mixture of aniline, with an aniline a step higher, prepared from toluene. If I were to give you the formula of Magenta you would be astonished at its complexity and size, but I think now you will see that it is really built up of aniline derivatives. Methyl Violet is a colour we have already referred to, and its chemical structure is still more complex, but it also is built up of aniline materials, and so is a basic aniline colour. Now it is possible for the colour-maker to prepare a very fine green dye from this beautiful violet (Methyl Violet). In fact he may convert the violet into the green colour by heating the first under pressure with a gas called methyl chloride (CH_3Cl). Methyl Violet is constructed of aniline or substituted aniline groups; the addition of CH_3Cl, then, gives us the Methyl Green. But one of the misfortunes of Methyl Green is that if the fabric dyed with it be boiled with water, at that temperature (212° F.) [Pg 97] the colour is decomposed and injured, for some of the methyl chloride in the compound is driven off. In fact, by stronger heating we may drive off all the methyl chloride and get the original Methyl Violet back again.

But we have coal-tar colours which are not basic, but rather of the nature of acid,—a better term would be *phenolic*, or of the nature of phenol or carbolic acid. Let us see what phenol or carbolic acid is. We saw that water may be formulated HOH, and that benzene is C_6H_6. Well, carbolic acid or phenol is a derivative of water, or a derivative of benzene, just as you like, and it is formulated C_6H_5OH. You can easily prove this by dropping carbolic acid or phenol down a red-hot tube filled with iron-borings. The oxygen is taken up by the iron to give oxide of iron, and benzene is obtained, thus: C_6H_5OH gives O and C_6H_6. But there is another hydrocarbon called naphthalene, $C_{10}H_8$, and this forms not one, but two phenols. As the name of the hydrocarbon is naphthalene, however, we call these compounds naphthols, and one is distinguished as alpha- the other as beta-naphthol, both of them having the formula $C_{10}H_7OH$. But now with respect to the colours. If we treat phenol with nitric acid under proper conditions, we get a yellow dye called picric acid, which is trinitro-phenol $C_6H_2(NO_2)_3OH$; you see this is no aniline dye; it is not a basic colour, for it would saturate, *i.e.* destroy the basicity of bases. Again, by oxidising phenol with oxalic acid and vitriol, we get a colour dyeing silk orange, namely, Aurin, $HO.C[C_6H_4(OH)]_3$. This is also an acid or phenolic dye, as a glance at its formula will show you. Its compound atom bristles, so to say, with phenol-residues, as some of the aniline dyes do with aniline residue-groups.

We come now to a peculiar but immensely important group of colours known as the azo-dyes, and these can be basic or acid, or of mixed kind. Just suppose two ammonia groups, NH_3 and NH_3. If we rob those nitrogen atoms of their [Pg 98] hydrogen atoms, we should leave two unsatisfied nitrogen atoms, atoms with an exceedingly keen appetite represented in terms of hydrogen atoms as N≡ and N≡. We might suppose a group, though of two N atoms partially satisfied by partial union with each other, thus—N:N—. Now this group forms the nucleus of the azo-colours, and if we satisfy a nitrogen at one side with an aniline, and at the other with a phenol, or at both ends with anilines, and so on, we get azo-dyes produced. The number of coal-tar colours is thus very great, and the variety also.

Adjective Colours.—As regards the artificial coal-tar adjective dye-stuffs, the principal are Alizarin and Purpurin. These are now almost entirely prepared from coal-tar anthracene, and madder and garancine are almost things of the past. Vegetable adjective colours are Brazil wood, containing the dye-generating principle Brasilin, logwood, containing Hæmatein, and santal-wood, camwood, and barwood, containing Santalin. Animal adjective colours are cochineal and lac dye. Then of wood colours we have further: quercitron, Persian berries, fustic and the tannins or tannic acids, comprising extracts, barks, fruits, and gallnuts, with also leaves and twigs, as with sumac. All these colours dye only with mordants, mostly forming with certain metallic oxides or basic salts, brightly-coloured compounds on the tissues to which they are applied.

[Pg 99]

LECTURE XI

DYEING OF WOOL AND FUR; AND OPTICAL PROPERTIES OF COLOURS

You have no doubt a tolerably vivid recollection of the illustrations given in Lecture I., showing the structure of the fibre of wool and fur. We saw that the wool fibre, of which fur might be considered a coarser quality, possesses a peculiar, complex, scaly structure, the joints reminding one of the appearance of plants of the *Equisetum* family, whilst the scaled structure resembles that of the skin of the serpent. Now you may easily understand that a structure like this, if it is to be completely and uniformly permeated by a dye liquor or any other aqueous solution, must have those scales not only well opened, but well cleansed, because if choked with greasy or other foreign matter impervious to or resisting water, there can be no chance of the mordanting or dye liquids penetrating uniformly; the resulting dye must be of a patchy nature. All wool, in its natural state, contains a certain amount of a peculiar compound almost like a potash soap, a kind of soft soap, but it also contains besides, a kind of fatty substance united with lime, and of a more insoluble nature than the first. This natural greasy matter is termed "yolk" or "suint"; and it ought never to be thrown away, as it sometimes is by the wool-scourers in this country, for it contains a substance resembling a fat named *cholesterin* or *cholesterol*, which is of great therapeutical value. Water alone will wash out a considerable amount of [Pg 100] this greasy matter, forming a kind of lather with it, but not all. As is almost invariably the case, after death, the matters and secretions which in life favour the growth and development of the parts, then commence to do the opposite. It is as if the timepiece not merely comes to a standstill, but commences to run backwards. This natural grease, if it be allowed to stand in contact with the wool for some time after shearing, instead of nourishing and preserving the fibres as it does on the living animal, commences to ferment, and injures them by making them hard and brittle. We see, then, the importance of "scouring" wool for the removal of "yolk," as it is called, dirt, oil, etc. If this important operation were omitted, or incompletely carried out, each fibre would be more or less covered or varnished with greasy matter, resisting the absorption and fixing of mordant and dye. As scouring agents, ammonia, carbonate of ammonia, carbonate of soda completely free from caustic, and potash or soda soaps, especially palm-oil soaps, which need not be made with bleached palm oil, but which must be quite free from free alkali, may be used. In making these palm-oil soaps it is better to err on the side of a little excess of free oil or fat, but if more than 1 per cent. of free fat be present, lathering qualities are then interfered with. Oleic acid soaps are excellent, but are rather expensive for wool; they are generally used for silks. Either as a skin soap or a soap for scouring wools, I should prefer one containing about ½ per cent. of free fatty matter, of course perfectly equally distributed, and not due to irregular saponification. On the average the soap solution for scouring wool may contain about 6½ oz. of soap to the gallon of water. In order to increase the cleansing powers of the soap solution, some ammonia may be added to it. However, if soap is used for wool-scouring, one thing must be borne in mind, namely, that the water used must not be hard, for if insoluble lime and magnesia soaps are formed and precipitated on the fibre, the scouring will [Pg 101] have removed one evil, but replaced it by another. The principal scouring material used is one of the various forms of commercial carbonate of soda, either alone or in conjunction with soap. Whatever be the form or name under which the carbonate of soda is sold, it must be free from hydrate of soda, *i.e.* caustic soda, or, as it is also termed, "causticity." By using this carbonate of soda you may dispense with soap, and so be able, even with a hard or calcareous water, to do your wool-scouring without anything like the ill effects that follow the use of soap and calcareous water. The carbonate of soda solutions ought not to exceed the specific gravity of 1° to 2° Twaddell (1½ to 3 oz. avoird. per gallon of water). The safest plan is to work with as considerable a degree of dilution and as low a temperature as are consistent with fetching the dirt and grease off. The scouring of loose wool, as we may now readily discern, divides itself into three stages: 1st, the stage in which those "yolk" or "suint" constituents soluble in water, are removed by steeping and washing in water. This operation is generally carried out by the wool-grower himself, for he desires to sell wool, and not wool plus "yolk" or "suint," and thus he saves himself considerable cost in transport. The water used in this process should not be at a higher temperature than 113° F., and the apparatus ought to be provided with an agitator; 2nd, the cleansing or scouring proper, with a weak alkaline solution; 3rd, the rinsing or final washing in water.

Thus far we have proceeded along the same lines as the woollen manufacturer, but now we must deviate from that course, for he requires softness and delicacy for special purposes, for spinning and weaving, etc.; but the felt manufacturer, and especially the manufacturer of felt for felt hats, requires to sacrifice some of this softness and delicacy in favour of greater felting powers, which can only be obtained by raising the scales of the fibres by means of a suitable process, such as [Pg 102] treatment with acids. This process is one which is by no means unfavourable to the dyeing capacities of the wool; on the whole it is decidedly favourable.

So far everything in the treatment of the wool has been perfectly favourable for the subsequent operations of the felt-hat dyer, but now I come to a process which I consider I should be perfectly unwarranted in passing over before proceeding to the dyeing processes. In fact, were it not for this "proofing process" (see Lecture VII.) the dyeing of felt hats would be as simple and easy of attainment as the ordinary dyeing of whole-wool fabrics. Instead of this, however, I consider the hat manufacturer, as regards his dyeing processes as applied to the stiffer classes of felt hats, has difficulties to contend with fully comparable with those which present themselves to the dyer of mixed cotton and woollen or Bradford goods. You have heard that the purpose of the wool-scourer is to remove the dirt, grease, and so-called yolk, filling the pores and varnishing the fibres. Now the effect of the work of the felt or felt-hat proofer is to undo nearly all this for the sake of rendering the felt waterproof and stiff. The material used, also, is even more impervious and resisting to the action of aqueous solutions of dyes and mordants than the raw wool would be. In short, it is impossible to mordant and to dye shellac by any process that will dye wool. To give you an idea of what it is necessary to do in order to colour or dye shellac, take the case of coloured sealing-wax, which is mainly composed of shellac, four parts, and Venice turpentine, one part. To make red sealing-wax this mixture is melted, and three parts of vermilion, an insoluble metallic pigment, are stirred in. If black sealing-wax is required, lamp-black or ivory-black is stirred in. The fused material is then cast in moulds, from which the sticks are removed on cooling. That is how shellac may be coloured as sealing-wax, but it is a totally different method from that by which wool is dyed. The difficulty then is this — in proofing, [Pg 103] your hat-forms are rendered impervious to the dye solutions of your dye-baths, all except a thin superficial layer, which then has to be rubbed down, polished, and finished. Thus in a short time, since the bulk of that superficially dyed wool or fur on the top of every hat is but small, and has been much reduced by polishing and rubbing, you soon hear of an appearance of bareness — I was going to say threadbareness — making itself manifest. This is simply because the colour or dye only penetrates a very little way down into the substance of the felt, until, in fact, it meets the proofing, which, being as it ought to be, a waterproofing, cannot be dyed. It cannot be dyed either by English or German methods; neither logwood black nor coal-tar blacks can make any really good impression on it. Cases have often been described to me illustrating the difficulty in preventing hats which have been dyed black with logwood, and which are at first a handsome deep black, becoming rather too soon of a rusty or brownish shade. Now my belief is that two causes may be found for this deterioration. One is the unscientific method adopted in many works of using the same bath practically for about a month together without complete renewal. During this time a large quantity of a muddy precipitate accumulates, rich in hydrated oxide of iron or basic iron salts of an insoluble kind. This mud amounts to no less than 25 per cent. of the weight of the copperas used. From time to time carbonate of ammonia is added to the bath, as it is said to throw up "dirt." The stuff or "dirt," chiefly an ochre-like mass stained black with the dye, and rich in iron and carbonate of iron, is skimmed off, and fresh verdigris and copperas added with another lot of hat-forms. No doubt on adding fresh copperas further precipitation of iron will take place, and so this ochre-like precipitate will accumulate, and will eventually come upon the hats like a kind of thin black mud. Now the effect of this will be that the dyestuff, partly in the fibre as a proper dye, and not [Pg 104] a little on the fibre as if "smudged" on or painted on, will, on exposure to the weather, moisture, air, and so on, gradually oxidise, the great preponderance of iron on the fibre changing to a kind of iron-rust, corroding the fibres in the process, and thus at once accounting for the change to the ugly brownish shade, and to the rubbing off and rapid wearing away of the already too thin superficial coating of dyed felt fibre. In the final spells of dyeing in the dye-beck already referred to, tolerably thick with black precipitate or mud, the application of black to the hat-forms begins, I fear, to assume at length a too close analogy to another blacking process closely associated with a pair of brushes and the time-honoured name of Day & Martin. With that logwood black fibre, anyone could argue as to a considerable proportion of the dye rubbing, wearing, or washing off. Thus, then, we have the second cause of the deterioration of the black, for the colour could not go into the fibre, and so it was chiefly laid or plastered on. You can also see that a logwood black hat dyer may well make the boast, and with considerable appearance of truth, that for the purposes of the English hat manufacturers, logwood black dyeing is the most appropriate, *i.e.* for the dyeing of highly proofed and stiff goods, but as to the permanent character of the black colour on those stiff hats, there you have quite another question. I firmly believe that in order to get the best results either with logwood black or "aniline blacks," it is absolutely necessary to have in possession a more scientific and manageable process of proofing. Such a process is that invented by F.W. Cheetham (see Lecture VII. p. 66).

In the dyeing of wool and felt with coal-tar colours, it is in many cases sufficient to add the solution of the colouring matters to the cold or tepid water of the dye-bath, and, after introducing the woollen material, to raise the temperature of the bath. The bath is generally heated to the boiling-point, and kept there for some time. A large number of these coal-tar colours [Pg 105] show a tendency of going so rapidly and greedily on to the fibre that it is necessary to find means to restrain them. This is done by adding a certain amount of Glauber's salts (sulphate of soda), in the solution of which coal-tar colours are not so soluble as in water alone, and so go more slowly, deliberately, and thus evenly upon the fibre. It is usually also best to dye in a bath slightly acid with sulphuric acid, or to add some bisulphate of soda. There is another point that needs good heed taking to, namely, in using different coal-tar colours to produce some mixed effect, or give some special shade, the colours to be so mixed must possess compatibility under like circumstances. For example, if you want a violet of a very blue shade, and you take Methyl Violet and dissolve it in water and then add Aniline Blue also in solution, you find that precipitation of the colour takes place in flocks. A colouring matter which requires, as some do, to be applied in an acid bath, ought not to be applied simultaneously with one that dyes best in a neutral bath. Numerous descriptions of methods of using coal-tar dyestuffs in hat-dyeing are available in different volumes of the *Journal of the Society of Chemical Industry*, and also tables for the detection of such dyestuffs on the fibre.

Now I will mention a process for dyeing felt a deep dead black with a coal-tar black dye which alone would not give a deep pure black, but one with a bluish-purple shade. To neutralise this purple effect, a small quantity of a yellow dyestuff and a trifle of indigotin are added. A deep black is thus produced, faster to light than logwood black it is stated, and one that goes on the fibre with the greatest ease. But I have referred to the use of small quantities of differently coloured dyes for the purpose of neutralising or destroying certain shades in the predominating colour. Now I am conscious that this matter is one that is wrapped in complete mystery, and far from the true ken of many of our dyers; but the rational treatment of such questions possesses such vast advantages, and pre-supposes [Pg 106] a certain knowledge of the theory of colour, of application and advantage so equally important, that I am persuaded I should not close this course wisely without saying a few words on that subject, namely, the optical properties of colours.

Colour is merely an impression produced upon the retina, and therefore on the brain, by various surfaces or media when light falls upon them or passes through them. Remove the light, and colour ceases to exist. The colour of a substance does not depend so much on the chemical character of that substance, but rather and more directly upon the physical condition of the surface or medium upon which the light falls or through which it passes. I can illustrate this easily. For example, there is a bright-red paint known as Crooke's heat-indicating paint. If a piece of iron coated with this paint be heated to about 150° F., the paint at once turns chocolate brown, but it is the same chemical substance, for on cooling we get the colour back again, and this can be repeated any number of times. Thus we see that it is the peculiar physical structure of bodies which appear coloured that has a certain effect upon the light, and hence it must be from the light itself that colour really emanates. Originally all colour proceeds from the source of light, though it seems to come to the eye from the apparently coloured objects. But without some elucidation this statement would appear as an enigma, since it might be urged that the light of the sun as well as that of artificial light is white, and not coloured. I hope, however, to show you that that light is white, because it is so much coloured, so variously and evenly coloured, though I admit the term "coloured" here is used in a special sense. White light contains and is made up of all the differently coloured rainbow rays, which are continually vibrating, and whose wave-lengths and number of vibrations distinguish them from each other. We will take some white light from an electric lantern and throw it on a screen. In a prism of glass we have a simple instrument for unravelling those rays, and

[Pg 107]

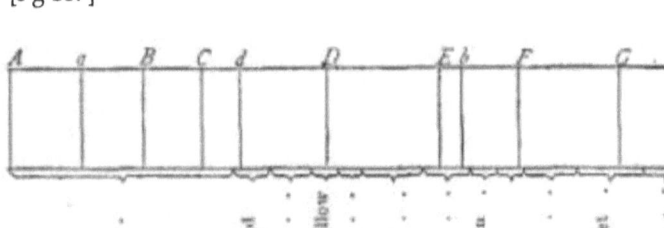

Fig. 16.

instead of letting them all fall on the same spot and illumine it with a white light, it causes them to fall side by side; in fact they all fall apart, and the prism has actually analysed that light. We get now a coloured band, similar to that of the rainbow, and this band is called the spectrum (see Fig. 16). If we could now run all these coloured rays together again, we should simply reproduce white light. We can do this by catching the coloured band in another prism, when the light now emerging will be found to be white. Every part of that spectrum consists of homogeneous light, *i.e.* light that cannot be further split up. The way in which the white light is so unravelled by the prism is this: As the light passes through the prism its different component coloured rays are variously deflected from their normal course, so that on emerging we have each of these coloured rays travelling in its own direction, vibrating in its own plane. It is well to remember that the bending off, or deflection, or refraction, is towards the thick end of the prism always, and that those of the coloured rays in that analysed band, the spectrum, most bent away from the original line of direction of the white light striking the prism, are said to be the most refrangible rays, and consequently are situated in the most refrangible end or part of the spectrum, namely, that farthest from the original direction [Pg 108] of the incident white light. These most refrangible rays are the violet, and we pass on to the least refrangible end, the red, through bluish-violet, blue, bluish-green, green, greenish-yellow, yellow, and orange. If you placed a prism say in the red part of the spectrum, and caught some of those red rays and allowed them to pass through your prism, and then either looked at the emerging light or let it fall on a white surface, you would find only red light would come through, only red rays. That light has been once analysed, and it cannot be further broken up. There is great diversity of shades, but only a limited number of primary impressions. Of these primary impressions there are only four—red, yellow, green, and blue, together with white and black. White is a collective effect, whilst black is the antithesis of white and the very negation of colour. The first four are called primary colours, for no human eye ever detected in them two different colours, while all of the other colours contain two or more primary colours. If we mix the following tints of the spectrum, *i.e.* the following rays of coloured light, we shall produce white light, red and greenish-yellow, orange and Prussian blue, yellow and indigo blue, greenish-yellow and violet. All those pairs of colours that unite to produce white are termed complementary colours. That is, one is complementary to the other. Thus if in white light you suppress any one coloured strip of rays, which, mingled uniformly with all the rest of the spectral rays, produces the white light, then that light no longer remains white, but is tinged with some particular tint. Whatever colour is thus suppressed, a particular other tint then pervades the residual light, and tinges it. That tint which thus makes its appearance is the one which, with the colour that was suppressed, gave white light, and the one is complementary to the other. Thus white can always be compounded of two tints, and these two tints are complementary colours. But it is important to remark here that I am now speaking of rays of coloured light proceeding [Pg 109] to and striking the eye; for a question like this might be asked: "You say that blue and yellow are complementary colours, and together they produce white, but if we mix a yellow and a blue paint or dye we have as the result a green colour. How is this?" The cases are entirely different, as I shall proceed to show. In speaking of the first, the complementary colours, we speak of pure spectral colours, coloured rays of light; in the latter, of pigment or dye colours. As we shall see, in the first, we have an addition direct of coloured lights producing white; in the latter, the green colour, appearing as the result of the mixture of the blue and yellow pigments, is obtained by the subtraction of colours; it is due to the absorption, by the blue and yellow pigments, of all the spectrum, practically, except the green portion. In the case of coloured objects, we are then confronted with the fact that these objects appear coloured because of an absorption by the colouring matter of every part of the rays of light falling thereupon, except that of the colour of the object, which colour is thrown off or reflected. This will appear clearer as we proceed. Now let me point out a further fact and indicate another step which will show you the value of such knowledge as this if properly applied. I said that if we selected from the coloured light spectrum, separated from white light by a prism, say, the orange portion, and boring a hole in our screen, if we caught that orange light in another prism, it would emerge as orange light, and suffer no further analysis. It cannot be resolved into red and yellow, as some might have supposed, it is monochromatic light, *i.e.* light purely of one colour. But when a mixture of red and yellow light, which means, of course, a mixture of rays of greater and less refrangibility respectively than our spectral orange, the monochromatic orange — is allowed to strike the eye, then we have again the impression of orange. How are we to distinguish a pure and monochromatic orange colour from a colour produced by a mixture of red and yellow? [Pg 110] In short, how are we to distinguish whether colours are homogeneous or mixed? The answer is, that this can only be done by the prism, apart from chemical analysis or testing of the substances.

The spectroscope is a convenient prism-arrangement, such that the analytical effect produced by that prism is looked at through a telescope, and the light that falls on the prism is carefully preserved from other light by passing it along a tube after only admitting a small quantity through a regulated slit.

Now all solid and liquid bodies when raised to a white heat give a continuous spectrum, one like the prismatic band already described, and one not interrupted by any dark lines or bands. The rays emitted from the white-hot substance of the sun have to pass, before reaching our earth, through the sun's atmosphere, and since the light emitted from any incandescent body is absorbed on passing through the vapour of that substance, and since the sun is surrounded by such an atmosphere of the vapours of various metals and substances, hence we have, on examining the sun's spectrum, instead of coloured bands or lines only, many dark ones amongst them, which are called Fraunhofer's lines. Ordinary incandescent vapours from highly heated substances give discontinuous spectra, *i.e.* spectra in which the rays of coloured light are quite limited, and they appear in the spectroscope only as lines of the breadth of the slit. These are called line-spectra, and every chemical element possesses in the incandescent gaseous state its own characteristic lines of certain colour and certain refrangibility, by means of which that element can be recognised. To observe this you place a Bunsen burner opposite the slit of the spectroscope, and introduce into its colourless flame on the end of a platinum wire a little of a volatile salt of the metal or element to be examined. The flame of the lamp itself is often coloured with a distinctiveness that is sufficient for a judgment to be made with the aid of the naked eye alone, as to the metal or element [Pg 111] present. Thus soda and its salts give a yellow flame, which is absolutely yellow or monochromatic, and if you look through your prism or spectroscope at it, you do not see a coloured rainbow band or spectrum, as with daylight or gaslight, but only one yellow double line, just where the yellow would have been if the whole spectrum had been represented. I think it is now plain that for the sake of observations and exact discrimination, it is necessary to map out our spectrum, and accordingly, in one of the tubes, the third, the spectroscope is provided with a graduated scale, so adjusted that when we look at the spectrum we also see the graduations of the scale, and so our spectrum is mapped; the lines marked out and named with the large and small letters of the alphabet, are certain of the prominent Fraunhofer's lines (see A, B, C, a, d, etc., Fig. 16). We speak, for example, of the soda yellow-line as coinciding with D of the spectrum. These, then, are spectra produced by luminous bodies.

The colouring matters and dyes, their solutions, and the substances dyed with them, are not, of course, luminous, but they do convert white light which strikes upon or traverses them into coloured light, and that is why they, in fact, appear either as coloured substances or solutions. The explanation of the coloured appearance is that the coloured substances or solutions have the power to absorb from the white light that strikes or traverses them, all the rays of the spectrum but those which are of the colour of the substance or solution in question, these latter being thrown off or reflected, and so striking the eye of the observer. Take a solution of Magenta, for example, and place a light behind it. All the rays of that white light are absorbed except the red ones, which pass through and are seen. Thus the liquid appears red. If a dyed piece be taken, the light strikes it, and if a pure red, from that light all the rays but red are absorbed, and so red light alone is reflected from its surface. But [Pg 112] this is not all with a dyed fabric, for here the light is not simply reflected light; part of it has traversed the upper layers of that coloured body, and is then reflected from the interior, losing a portion of its coloured rays by absorption. This reflected coloured light is always mixed with a certain amount of white light reflected from the actual surface of the body before penetrating its uppermost layer. Thus, if dyed fabrics are examined by the spectroscope, the same appearances are generally observed as with the solution of the corresponding colouring matters. An absorption spectrum is in each case obtained, but the one from the solution is the purer, for it does not contain the mixed white light reflected from the surfaces of coloured objects. Let us now take an example. We will take a cylinder glass full of picric acid in water, and of a yellow colour. Now when I pass white light through that solution and examine the emerging light, which looks, to my naked eye, yellow, I find by the spectroscope that what has taken place is this: the blue part of the spectrum is totally extinguished as far as G and 2/3 of F. That is all.

Then why, say you, does that liquid look yellow if all the rest of those rays pass through and enter the eye, namely, the blue-green with a trifle of blue, the green, yellow, orange, and red? The reason is this: we have already seen that the colours complementary to, and so producing white light with red, are green and greenish-blue or bluish-green. Hence these cancel, so to say, and we only see yellow. We do not see a pure yellow, then, in picric acid, but yellow with a considerable amount of white. Here is a piece of scarlet paper. Why does it appear scarlet? Because from the white light falling upon it, it practically absorbs all the rays of the spectrum except the red and orange ones, and these it reflects. If this be so, then, and we take our spectrum band of perfectly pure colours and pass our strip of scarlet paper along that variously coloured band of light, we shall be able to test the truth of several statements I have [Pg 113] made as to the nature of colour. I have said colour is only an impression, and not a reality; and that it does not exist apart from light. Now, I can show you more, namely, that the colour of the so-called coloured object is entirely dependent on the existence in the light of the special coloured rays which it radiates, and that this scarlet paper depends on the red light of the spectrum for the existence of its redness. On passing the piece of scarlet paper along the coloured band of light, it appears red only when in the red portion of the spectrum, whilst in the other portions, though it is illumined, yet it has no colour, in fact it looks black. Hence what I have said is true, and, moreover, that red paper looks red because, as you see, it absorbs and extinguishes all the rays of the spectrum but the red ones, and these it radiates. A bright green strip of paper placed in the red has no colour, and looks black, but transferred to the pure green portion it radiates that at once, does not absorb it as it did the red, and so the green shines out finely. I have told you that sodium salts give to a colourless flame a fine monochromatic or pure yellow colour. Now, if this be so, and if all the light available in this world were of such a character, then such a colour as blue would be unknown. We will now ask ourselves another question, "We have a new blue colouring matter, and we desire to know if we may expect it to be one of the greatest possible brilliancy, what spectroscopic conditions ought it to fulfil?" On examining a solution of it, or rather the light passing through a solution of it, with the spectroscope, we ought to find that all the rays of the spectrum lying between and nearly to H and b (Fig. 16), *i.e.* all the bluish-violet, blue, and blue-green rays pass through it unchanged, unabsorbed, whilst all the rest should be completely absorbed. In like manner a pure yellow colour would allow all the rays lying between orange-red and greenish-yellow (Fig. 16) to pass through unchanged, but would absorb all the other colours of the spectrum.

[Pg 114]

Now we come to the, for you, most-important subject of mixtures of colours and their effects. Let us take the popular case of blue and yellow producing green. We have seen that the subjective effect of the mixture of blue and yellow light on the eye is for the latter to lose sense of colour, since colour disappears, and we get what we term white light; in strict analogy to this the objective effect of a pure yellow pigment and a blue is also to destroy colour, and so no colour comes from the object to the eye; that object appears black. Now the pure blue colouring matter would not yield a green with the pure yellow colouring matter, for if you plot off the two absorption spectra as previously described, on to the spectrum (Fig. 16), you will find that all the rays would be absorbed by the mixture, and the result would be a black. But, now, suppose a little less pure yellow were taken, one containing a little greenish-yellow and a trifle of green, and also a little orange-red on the other side to red, then whereas to the eye that yellow might be as good as the first; now, when mixed with a blue, we get a very respectable green. But, and this is very important, although of the most brilliant dyes and colours there are probably no two of these that would so unite to block out all the rays and produce black, yet this result can easily and practically be arrived at by using three colouring matters, which must be as different as possible from one another. Thus a combination of a red, a yellow, and a blue colouring matter, when concentrated enough, will not let any light pass through it, and can thus be used for the production of blacks, and this property is made use of in dyeing. And now we see why a little yellow dye is added to our coal-tar black. A purplish shade would else be produced; the yellow used is a colour complementary to that purple, and it absorbs just those blue and purple rays of the spectrum necessary to illuminate by radiation that purple, and *vice versâ*; both yellow and purple therefore disappear. In like manner, had the black been of

[Pg 115] a greenish shade, I should have added Croceine Orange, which on the fabric would absorb just those green and bluish rays of light necessary to radiate from and illumine that greenish part, and the greenish part would do the like by the orange rays; the effects would be neutralised, and all would fall together into black.

THE END.

[Pg 116]

FULL PARTICULARS OF CONTENTS

Of the Books mentioned in this **ABRIDGED CATALOGUE** will be found in the following Catalogues of

CURRENT TECHNICAL BOOKS.

LIST I.

Artists' Colours—Bone Products—Butter and Margarine Manufacture—Casein—Cements—Chemical Works (Designing and Erection)—Chemistry (Agricultural, Industrial, Practical and Theoretical)—Colour Mixing—Colour Manufacture—Compounding Oils—Decorating—Driers—Drying Oils—Drysaltery—Emery—Essential Oils—Fats (Animal, Vegetable, Edible)—Gelatines—Glues—Greases—Gums—Inks—Lead—Leather—Lubricants—Oils—Oil Crushing—Paints—Paint Manufacturing—Paint Material Testing—Perfumes—Petroleum—Pharmacy—Recipes (Paint, Oil and Colour)—Resins—Sealing Waxes—Shoe Polishes—Soap Manufacture—Solvents—Spirit Varnishes—Varnishes—White Lead—Workshop Wrinkles.

LIST II.

Bleaching—Bookbinding—Carpet Yarn Printing—Colour (Matching, Mixing, Theory)—Cotton Combing Machines—Dyeing (Cotton, Woollen and Silk Goods)—Dyers' Materials—Dye-stuffs—Engraving—Flax, Hemp and Jute Spinning and Twisting—Gutta-Percha—Hat Manufacturing—India-rubber—Inks—Lace-making—Lithography—Needlework—Paper Making—Paper-Mill Chemist—Paper-pulp Dyeing—Point Lace—Power-loom Weaving—Printing Inks—Silk Throwing—Smoke Prevention—Soaps—Spinning—Textile (Spinning, Designing, Dyeing, Weaving, Finishing)—Textile Materials—Textile Fabrics—Textile Fibres—Textile Oils—Textile Soaps—Timber—Water (Industrial Uses)—Water-proofing—Weaving—Writing Inks—Yarns (Testing, Sizing).

LIST III.

Architectural Terms—Brassware (Bronzing, Burnishing, Dipping, Lacquering)—Brickmaking—Building—Cement Work—Ceramic Industries—China—Coal-dust Firing—Colliery Books—Concrete—Condensing Apparatus—Dental Metallurgy—Drainage—Drugs—Dyeing—Earthenware—Electrical Books—Enamelling—Enamels—Engineering Handbooks—Evaporating Apparatus—Flint Glassmaking—Foods—Food Preserving—Fruit Preserving—Gas Engines—Gas Firing—Gearing—Glassware (Painting, Riveting)—Hops—Iron (Construction, Science)—Japanning—Lead—Meat Preserving—Mines (Haulage, Electrical Equipment, Ventilation, Recovery Work from)—Plants (Diseases, Fungicides, Insecticides)—Plumbing Books—Pottery (Architectural, Clays, Decorating, Manufacture, Marks on)—Reinforced Concrete—Riveting (China, Earthenware, Glassware)—Steam Turbines—Sanitary Engineering—Steel (Hardening, Tempering)—Sugar—Sweetmeats—Toothed Gearing—Vegetable Preserving—Wood Dyeing—X-Ray Work.

COPIES OF ANY OF THESE LISTS WILL BE SENT POST FREE ON APPLICATION.

(Paints, Colours, Pigments and Printing Inks.)

THE CHEMISTRY OF PIGMENTS. By Ernest J. Parry, B.Sc. (Lond.), F.I.C., F.C.S., and J.H. Coste, F.I.C., F.C.S. Demy 8vo. Five Illustrations. 285 pp. Price 10s. 6d. net. (Post free, 10s. 10d. home; 11s. 3d. abroad.)

THE MANUFACTURE OF PAINT. A Practical Handbook for Paint Manufacturers, Merchants and Painters. By J. Cruickshank Smith, B.Sc. Demy 8vo. 200 pp. Sixty Illustrations and One Large Diagram. Price 7s. 6d. net. (Post free, 7s. 10d. home; 8s. abroad.)

DICTIONARY OF CHEMICALS AND RAW PRODUCTS USED IN THE MANUFACTURE OF PAINTS, COLOURS, VARNISHES AND ALLIED PREPARATIONS. By George H. Hurst, F.C.S. Demy 8vo. 380 pp. Price 7s. 6d. net. (Post free, 8s. home; 8s. 6d. abroad.)

THE MANUFACTURE OF LAKE PIGMENTS FROM ARTIFICIAL COLOURS. By Francis H. Jennison, F.I.C., F.C.S. **Sixteen Coloured Plates, showing Specimens of Eighty-nine Colours, specially prepared from the Recipes given in the Book.** 136 pp. Demy 8vo. Price 7s. 6d. net. (Post free, 7s. 10d. home; 8s. abroad.)

THE MANUFACTURE OF MINERAL AND LAKE PIGMENTS. Containing Directions for the Manufacture of all Artificial, Artists and Painters' Colours, Enamel, Soot and Metallic Pigments. A textbook for Manufacturers, Merchants, Artists and Painters, By Dr. Josef Bersch. Translated by A.C. Wright, M.A. (Oxon.), B.Sc. (Lond.). Forty-three Illustrations. 476 pp. Demy 8vo. Price 12s. 6d. net. (Post free, 13s. home; 13s. 6d. abroad.)

RECIPES FOR THE COLOUR, PAINT, VARNISH, OIL, SOAP AND DRYSALTERY TRADES. Compiled by An Analytical Chemist. 350 pp. Second Revised Edition. Demy 8vo. Price 10s. 6d. net. (Post free, 11s. home; 11s. 3d. abroad.)

OIL COLOURS AND PRINTERS' INKS. By Louis Edgar Andés. Translated from the German. 215 pp. Crown 8vo. 56 Illustrations. Price 5s. net. (Post free, 5s. 4d. home; 5s. 6d. abroad.)

[Pg 4]

MODERN PRINTING INKS. A Practical Handbook for Printing Ink Manufacturers and Printers. By Alfred Seymour. Demy 8vo. Six Illustrations. 90 pages. Price 5s. net. (Post free, 5s. 4d. home; 5s. 6d. abroad.)

THREE HUNDRED SHADES AND HOW TO MIX THEM. For Architects, Painters and Decorators. By A. Desaint, Artistic Interior Decorator of Paris. The book contains 100 folio Plates, measuring 12 in. by 7 in., each Plate containing specimens of three artistic shades. These shades are all numbered, and their composition and particulars for mixing are fully given at the beginning of the book. Each Plate is interleaved with grease-proof paper, and the volume is very artistically bound in art and linen with the Shield of the Painters' Guild impressed on the cover in gold and silver. Price 21s. net. (Post free, 21s. 6d. home; 22s. 6d. abroad.)

HOUSE DECORATING AND PAINTING. By W. Norman Brown. Eighty-eight Illustrations. 150 pp. Crown 8vo. Price 3s. 6d. net. (Post free, 3s. 9d. home and abroad.)

A HISTORY OF DECORATIVE ART. By W. Norman Brown. Thirty-nine Illustrations. 96 pp. Crown 8vo. Price 1s. net. (Post free, 1s. 3d. home and abroad.)

WORKSHOP WRINKLES. for Decorators, Painters, Paperhangers, and Others. By W.N. Brown. Crown 8vo. 128 pp. Second Edition. Price 2s. 6d. net. (Post free, 2s. 9d. home; 2s. 10d. abroad.)

CASEIN. By Robert Scherer. Translated from the German by Chas. Salter. Demy 8vo. Illustrated. Second Revised English Edition. 160 pp. Price 7s. 6d. net. (Post free, 7s. 10d. home; 8s. abroad.)

SIMPLE METHODS FOR TESTING PAINTERS' MATERIALS. By A.C. Wright, M.A. (Oxon.)., B.Sc. (Lond.). Crown 8vo. 160 pp. Price 5s. net. (Post free, 5s. 3d. home; 5s. 6d. abroad.)

IRON-CORROSION, ANTI-FOULING AND ANTI-CORROSIVE PAINTS. Translated from the German of Louis Edgar Andés. Sixty-two Illustrations. 275 pp. Demy 8vo. Price 10s. 6d. net. (Post free, 10s. 10d. home; 11s. 3d. abroad.)

THE TESTING AND VALUATION OF RAW MATERIALS USED IN PAINT AND COLOUR MANUFACTURE. By M.W.

Jones, F.C.S. A Book for the Laboratories of Colour Works. 88 pp. Crown 8vo. Price 5s. net. (Post free, 5s. 3d. home and abroad.)

For contents of these books, see List I.

[Pg 5]

THE MANUFACTURE AND COMPARATIVE MERITS OF WHITE LEAD AND ZINC WHITE PAINTS. By G. Petit, Civil Engineer, etc. Translated from the French. Crown 8vo. 100 pp. Price 4s. net. (Post free, 4s. 3d. home; 4s. 4d. abroad.)

STUDENTS' HANDBOOK OF PAINTS, COLOURS, OILS AND VARNISHES. By John Furnell. Crown 8vo. 12 Illustrations. 96 pp. Price 2s. 6d. net. (Post free, 2s. 9d. home and abroad.)

(Varnishes and Drying Oils.)

THE MANUFACTURE OF VARNISHES AND KINDRED INDUSTRIES. By J. Geddes McIntosh. Second, greatly enlarged, English Edition, in three Volumes, based on and including the work of Ach. Livache.

Volume I.—**OIL CRUSHING, REFINING AND BOILING, THE MANUFACTURE OF LINOLEUM, PRINTING AND LITHOGRAPHIC INKS, AND INDIA-RUBBER SUBSTITUTES.** Demy 8vo. 150 pp. 29 Illustrations. Price 7s. 6d. net. (Post free, 7s. 10d. home; 8s. abroad.)

Volume II.—**VARNISH MATERIALS AND OIL-VARNISH MAKING.** Demy 8vo. 70 Illustrations. 220 pp. Price 10s. 6d. net. (Post free, 10s. 10d. home; 11s. 3d. abroad.)

Volume III.—**SPIRIT VARNISHES AND SPIRIT VARNISH MATERIALS.** Demy 8vo. Illustrated. 464 pp. Price 12s. 6d. net. (Post free, 13s. home; 13s. 6d. abroad.)

DRYING OILS, BOILED OIL AND SOLID AND LIQUID DRIERS. By L.E. Andés. Expressly Written for this Series of Special Technical Books, and the Publishers hold the Copyright for English and Foreign Editions. Forty-two Illustrations. 342 pp. Demy 8vo. Price 12s. 6d. net. (Post free, 13s. home; 13s. 3d. abroad.)

(Analysis of Resins, see page 9.)

(Oils, Fats, Waxes, Greases, Petroleum.)

LUBRICATING OILS, PATS AND GREASES: Their Origin, Preparation, Properties, Uses and Analyses. A Handbook for Oil Manufacturers, Refiners and Merchants, and the Oil and Fat Industry in General. By George H. Hurst, F.C.S. Third Revised and Enlarged Edition. Seventy-four Illustrations. 384 pp. Demy 8vo. Price 10s. 6d. net. (Post free, 11s. home; 11s. 3d. abroad.)

TECHNOLOGY OF PETROLEUM: Oil Fields of the World—Their History, Geography and Geology—Annual Production and Development—Oil-well Drilling—Transport. By Henry Neuberger and Henry Noalhat. Translated from the French by J.G. McIntosh. 550 pp. 153 Illustrations. 26 Plates. Super Royal 8vo. Price 21s. net. (Post free, 21s, 9d. home; 23s. 6d. abroad.)

MINERAL WAXES: Their Preparation and Uses. By Rudolf Gregorius. Translated from the German. Crown 8vo. 250 pp. 32 Illustrations. Price 6s. net. (Post free, 6s. 4d. home; 6s. 6d. abroad.)

THE PRACTICAL COMPOUNDING OF OILS, TALLOW AND GREASE FOR LUBRICATION, ETC. By An Expert Oil Refiner. Second Edition. 100 pp. Demy 8vo. Price 7s. 6d. net. (Post free, 7s. 10d. home; 8s. abroad.)

THE MANUFACTURE OF LUBRICANTS, SHOE POLISHES AND LEATHER DRESSINGS. By Richard Brunner. Translated from the Sixth German Edition by Chas. Salter. 10 Illustrations. Crown 8vo. 170 pp. Price 7s. 6d. net. (Post free, 7s. 10d. home; 8s. abroad.)

THE OIL MERCHANTS' MANUAL AND OIL TRADE READY RECKONER. Compiled by Frank F. Sherriff. Second Edition Revised and Enlarged. Demy 8vo. 214 pp. With Two Sheets of Tables. Price 7s. 6d. net. (Post free, 7s. 10d. home; 8s. 3d. abroad.)

ANIMAL FATS AND OILS: Their Practical Production, Purification and Uses for a great Variety of Purposes. Their Properties, Falsification and Examination. Translated from the German of Louis Edgar Andés. Sixty-two Illustrations. 240 pp. Second Edition, Revised and Enlarged. Demy 8vo., Price 10s. 6d. net. (Post free, 10s. 10d. home; 11s. 3d. abroad.)

For contents of these books, see List I.

[Pg 7]

VEGETABLE FATS AND OILS: Their Practical Preparation, Purification and Employment for Various Purposes, their Properties, Adulteration and Examination. Translated from the German of Louis Edgar Andés. Ninety-four Illustrations. 340 pp. Second Edition. Demy 8vo. Price 10s. 6d. net. (Post free, 11s. home; 11s. 6d. abroad.)

EDIBLE FATS AND OILS: Their Composition, Manufacture and Analysis. By W.H. Simmons, B.Sc. (Lond.), and C.A. Mitchell, B.A. (Oxon.). Demy 8vo. 150 pp. Price 7s. 6d. net. (Post free, 7s. 9d. home; 8s. abroad.)

(Essential Oils and Perfumes.)

THE CHEMISTRY OF ESSENTIAL OILS AND ARTIFICIAL PERFUMES. By Ernest J. Parry, B.Sc. (Lond.), F.I.C., F.C.S. Second Edition, Revised and Enlarged. 552 pp. 20 Illustrations. Demy 8vo. Price 12s. 6d. net. (Post free, 13s. home; 13s. 6d. abroad.)

(Soap Manufacture.)

SOAPS. A Practical Manual of the Manufacture of Domestic, Toilet and other Soaps. By George H. Hurst, F.C.S. 2nd edition. 390 pp. 66 Illustrations. Demy 8vo. Price 12s. 6d. net. (Post free, 13s. home; 13s. 6d. abroad.)

TEXTILE SOAPS AND OILS. Handbook on the Preparation, Properties and Analysis of the Soaps and Oils used in Textile Manufacturing, Dyeing and Printing. By George H. Hurst, F.C.S. Crown 8vo. 195 pp. 1904. Price 5s. net. (Post free, 5s. 4d. home; 5s. 6d. abroad.)

THE HANDBOOK OF SOAP MANUFACTURE. By Wm. H. Simmons, B.Sc. (Lond.), F.C.S. and H.A. Appleton. Demy 8vo. 160 pp. 27 Illustrations. Price 8s. 6d. net. (Post free, 8s. 10d. home; 9s. abroad.)

(Cosmetical Preparations.)

COSMETICS: MANUFACTURE, EMPLOYMENT AND TESTING OF ALL COSMETIC MATERIALS AND COSMETIC SPECIALITIES. Translated from the German of Dr. Theodor Koller. Crown 8vo. 262 pp. Price 5s. net. (Post free, 5s. 4d. home; 5s. 6d. abroad.)

[Pg 8]

(Glue, Bone Products and Manures.)

GLUE AND GLUE TESTING. By Samuel Rideal, D.Sc. (Lond.), F.I.C. Fourteen Engravings. 144 pp. Demy 8vo. Price 10s. 6d. net. (Post free, 10s. 10d. home; 11s. abroad)

BONE PRODUCTS AND MANURES: An Account of the most recent Improvements in the Manufacture of Fat, Glue, Animal Charcoal, Size, Gelatine and Manures. By Thomas Lambert, Technical and Consulting Chemist. Illustrated by Twenty-one Plans and Diagrams. 162 pp. Demy 8vo. Price 7s. 6d. net. (Post free, 7s. 10d. home; 8s. abroad.)

(*See also Chemical Manures, p. 9.*)

(Chemicals, Waste Products, etc.)

REISSUE OF **CHEMICAL ESSAYS OF C.W. SCHEELE.** First Published in English in 1786. Translated from the Academy of Sciences at Stockholm, with Additions. 300 pp. Demy 8vo. Price 5s. net. (Post free, 5s. 6d. home; 5s. 9d. abroad.)

THE MANUFACTURE OF ALUM AND THE SULPHATES AND OTHER SALTS OF ALUMINA AND IRON. Their Uses and Applications as Mordants in Dyeing and Calico Printing, and their other Applications in the Arts Manufactures, Sanitary Engineering, Agriculture and Horticulture. Translated from the French of Lucien Geschwind. 195 Illustrations. 400 pp. Royal 8vo. Price 12s. 6d. net. (Post free, 13s. home; 13s. 6d. abroad.)

AMMONIA AND ITS COMPOUNDS: Their Manufacture and Uses. By Camille Vincent, Professor at the Central School of Arts and Manufactures, Paris. Translated from the French by M.J. Salter.

Royal 8vo. 114 pp. Thirty-two Illustrations. Price 5s. net. (Post free, 5s. 4d. home; 5s. 6d. abroad.)

CHEMICAL WORKS: Their Design, Erection, and Equipment. By S.S. Dyson and S.S. Clarkson. Royal 8vo. 220 pp. With Plates and Illustrations. Price 21s. net. (Post free, 21s. 6d. home; 22s. abroad.)

SHALE TAR DISTILLATION: The Treatment of Shale and Lignite Products. Translated from the German of W. Scheithauer. [*In the Press.*

For contents of these books, see List I.

[Pg 9]

INDUSTRIAL ALCOHOL. A Practical Manual on the Production and Use of Alcohol for Industrial Purposes and for Use as a Heating Agent, as an Illuminant and as a Source of Motive Power. By J.G. McIntosh, Lecturer on Manufacture and Applications of Industrial Alcohol at The Polytechnic, Regent Street, London. Demy 8vo. 1907. 250 pp. With 75 Illustrations and 25 Tables. Price 7s. 6d. net. (Post free, 7s. 9d. home; 8s. abroad.)

THE UTILISATION OF WASTE PRODUCTS. A Treatise on the Rational Utilisation, Recovery and Treatment of Waste Products of all kinds. By Dr. Theodor Koller. Translated from the Second Revised German Edition. Twenty-two Illustrations. Demy 8vo. 280 pp. Price 7s. 6d. net. (Post free, 7s. 10d. home; 8s. 3d. abroad.)

ANALYSIS OF RESINS AND BALSAMS. Translated from the German of Dr. Karl Dieterich. Demy 8vo. 340 pp. Price 7s. 6d. net. (Post free, 7s. 10d. home; 8s. 3d. abroad.)

(Agricultural Chemistry and Manures.)

MANUAL OF AGRICULTURAL CHEMISTRY. By Herbert Ingle, F.I.C., Late Lecturer on Agricultural Chemistry, the Leeds University; Lecturer in the Victoria University. Second Edition, with additional matter relating to Tropical Agriculture, etc. 438 pp. 11 Illustrations. Demy 8vo. Price 7s. 6d. net. (Post free, 8s. home; 8s. 6d. abroad.)

CHEMICAL MANURES. Translated from the French of J. Fritsch. Demy 8vo. Illustrated. 340 pp. Price 10s. 6d. net. (Post free, 11s. home; 11s. 6d. abroad.)

(*See also Bone Products and Manures, p. 8.*)

(Writing Inks and Sealing Waxes.)

INK MANUFACTURE: Including Writing, Copying, Lithographic, Marking, Stamping, and Laundry Inks. By Sigmund Lehner. Three Illustrations. Crown 8vo. 162 pp. Translated from the German of the Fifth Edition. Price 5s. net. (Post free, 5s. 3d. home; 5s. 6d. abroad.)

SEALING-WAXES, WAFERS AND OTHER ADHESIVES FOR THE HOUSEHOLD, OFFICE, WORKSHOP AND FACTORY. By H.C. Standage, Crown 8vo. 96 pp. Price 5s. net. (Post free, 5s. 3d. home; 5s. 4d. abroad.)

[Pg 10]

(Lead Ores and Lead Compounds.)

LEAD AND ITS COMPOUNDS. By Thos. Lambert, Technical and Consulting Chemist. Demy 8vo. 226 pp. Forty Illustrations. Price 7s. 6d. net. (Post free, 7s. 10d. home; 8s. 3d. abroad.)

NOTES ON LEAD ORES: Their Distribution and Properties. By Jas. Fairie, F.G.S. Crown 8vo. 64 pages. Price 1s. net. (Post free, 1s. 3d. home; 1s. 4d. abroad.)

(*White Lead and Zinc White Paints, see p. 5. .*)

(Industrial Hygiene.)

THE RISKS AND DANGERS TO HEALTH OF VARIOUS OCCUPATIONS AND THEIR PREVENTION. By Leonard A. Parry, M.D., B.Sc. (Lond.). 196 pp. Demy 8vo. Price 7s. 6d. net. (Post free, 7s. 10d. home; 8s. abroad.)

(Industrial Uses of Air, Steam and Water.)

DRYING BY MEANS OF AIR AND STEAM. Explanations, Formulæ, and Tables for Use in Practice. Translated from the German of E. Hausbrand. Two folding Diagrams and Thirteen Tables. Crown 8vo. 72 pp. Price 5s. net. (Post free, 5s. 3d. home; 5s. 6d. abroad.)

(*See also "Evaporating, Condensing and Cooling Apparatus," p. 19.*)

PURE AIR, OZONE, AND WATER. A Practical Treatise of their Utilisation and Value in Oil, Grease, Soap, Paint, Glue and other Industries. By W.B. Cowell. Twelve Illustrations. Crown 8vo. 85 pp. Price 5s. net. (Post free, 5s. 3d. home; 5s. 6d. abroad.)

THE INDUSTRIAL USES OF WATER. COMPOSITION – EFFECTS – TROUBLES – REMEDIES – RESIDUARY WATERS – PURIFICATION – ANALYSIS. By H. de la Coux. Royal 8vo. Translated from the French and Revised by Arthur Morris. 364 pp. 135 Illustrations. Price 10s. 6d. net. (Post free, 11s. home; 11s. 6d. abroad.)

(*See Books on Smoke Prevention, Engineering and Metallurgy, p. 19, etc.*)

For contents of these books, see List III.

[Pg 11]

(X Rays.)

PRACTICAL X RAY WORK. By Frank T. Addyman, B.Sc. (Lond.), F.I.C., Member of the Roentgen Society of London; Radiographer to St. George's Hospital; Demonstrator of Physics and Chemistry, and Teacher of Radiography in St. George's Hospital Medical School. Demy 8vo. Twelve Plates from Photographs of X Ray Work. Fifty-two Illustrations. 200 pp. Price 10s. 6d. net. (Post free, 10s. 10d. home; 11s. 3d. abroad.)

(India-Rubber and Gutta Percha.)

INDIA-RUBBER AND GUTTA PERCHA. Second English Edition, Revised and Enlarged. Based on the French work of T. Seeligmann, G. Lamy Torrilhon and H. Falconnet by John Geddes McIntosh. Royal 8vo. 100 Illustrations. 400 pages. Price 12s. 6d. net. (Post free, 13s. home; 13s. 6d. abroad.)

(Leather Trades.)

THE LEATHER WORKER'S MANUAL. Being a Compendium of Practical Recipes and Working Formulæ for Curriers, Bootmakers, Leather Dressers, Blacking Manufacturers, Saddlers, Fancy Leather Workers. By H.C. Standage. Demy 8vo. 165 pp. Price 7s. 6d. net. (Post free, 7s. 10d. home; 8s. abroad.)

(See also Manufacture of Shoe Polishes, Leather Dressings, etc., p. 6.)

(Pottery, Bricks, Tiles, Glass, etc.)

MODERN BRICKMAKING. By Alfred B. Searle, Royal 8vo. 440 pages. 260 Illustrations. Price 12s. 6d. net. (Post free, 13s. home; 13s. 6d. abroad.)

THE MANUAL OF PRACTICAL POTTING. Compiled by Experts, and Edited by Chas. F. Binns. Third Edition, Revised and Enlarged. 200 pp. Demy 8vo. Price 17s. 6d. net. (Post free, 17s. 10d. home; 18s. 3d. abroad.)

POTTERY DECORATING. A Description of all the Processes for Decorating Pottery and Porcelain. By R. Hainbach. Translated from the German. Crown 8vo. 250 pp. Twenty-two Illustrations. Price 7s. 6d. net. (Post free, 7s. 10d. home; 8s. abroad.)

A TREATISE ON CERAMIC INDUSTRIES. A Complete Manual for Pottery, Tile, and Brick Manufacturers. By Emile Bourry. A Revised Translation from the French, with some Critical Notes by Alfred B. Searle. Demy 8vo. 308 Illustrations. 460 pp. Price 12s. 6d. net. (Post free, 13s. home; 13s. 6d. abroad.)

[Pg 12]

ARCHITECTURAL POTTERY. Bricks, Tiles, Pipes, Enamelled Terra-cottas, Ordinary and Incrusted Quarries, Stoneware Mosaics, Faïences and Architectural Stoneware. By Leon Lefêvre. Translated from the French by K.H. Bird, M.A., and W. Moore Binns. With Five Plates. 950 Illustrations in the Text, and numerous estimates. 500 pp., royal 8vo. Price 15s. net. (Post free, 15s. 6d. home; 16s. 6d. abroad.)

CERAMIC TECHNOLOGY: Being some Aspects of Technical Science as Applied to Pottery Manufacture. Edited by Charles F.

Binns. 100 pp. Demy 8vo. Price 12s. 6d. net. (Post free, 12s. 10d. home; 13s. abroad.)

THE ART OF RIVETING GLASS, CHINA AND EARTHENWARE. By J. Howarth. Second Edition. Paper Cover. Price 1s. net. (By post, home or abroad, 1s. 1d.)

NOTES ON POTTERY CLAYS. The Distribution, Properties, Uses and Analyses of Ball Clays, China Clays and China Stone. By Jas. Fairie, F.G.S. 132 pp. Crown 8vo. Price 3s. 6d. net. (Post free, 3s. 9d. home; 3s. 10d. abroad.)

HOW TO ANALYSE CLAY. By H.M. Ashby. Demy 8vo. 72 Pages. 20 Illustrations. Price 3s. 6d. net. (Post free, 3s. 9d. home; 3s. 10d. abroad.)

A Reissue of

THE HISTORY OF THE STAFFORDSHIRE POTTERIES; AND THE RISE AND PROGRESS OF THE MANUFACTURE OF POTTERY AND PORCELAIN. With References to Genuine Specimens, and Notices of Eminent Potters. By Simeon Shaw. (Originally published in 1829.) 265 pp. Demy 8vo. Price 5s. net. (Post free, 5s. 4d. home; 5s. 9d. abroad.)

A Reissue of

THE CHEMISTRY OF THE SEVERAL NATURAL AND ARTIFICIAL HETEROGENEOUS COMPOUNDS USED IN MANUFACTURING PORCELAIN, GLASS AND POTTERY. By Simeon Shaw. (Originally published in 1837.) 750 pp. Royal 8vo. Price 10s. net. (Post free, 10s. 6d. home; 12s. abroad.)

BRITISH POTTERY MARKS. By G. Woolliscroft Rhead. Demy 8vo. 310 pp. With over Twelve-hundred Illustrations of Marks. Price 7s. 6d. net. (Post free, 8s. home; 8s. 3d. abroad.)

For contents of these books, see List III.

[Pg 13]

(Glassware, Glass Staining and Painting.)

RECIPES FOR FLINT GLASS MAKING. By a British Glass Master and Mixer. Sixty Recipes. Being Leaves from the Mixing Book of several experts in the Flint Glass Trade, containing up-to-date recipes and valuable information as to Crystal, Demi-crystal and Coloured Glass in its many varieties. It contains the recipes for cheap metal suited to pressing, blowing, etc., as well as the most costly crystal and ruby. Second Edition. Crown 8vo. Price 10s. 6d. net. (Post free, 10s. 9d. home; 10s. 10d. abroad.)

A TREATISE ON THE ART OF GLASS PAINTING. Prefaced with a Review of Ancient Glass. By Ernest R. Suffling. With One Coloured Plate and Thirty-seven Illustrations. Demy 8vo. 140 pp. Price 7s. 6d. net. (Post free, 7s. 10d. home; 8s. abroad.)

(Paper Making, Paper Dyeing, and Testing.)

THE DYEING OF PAPER PULP. A Practical Treatise for the use of Papermakers, Paperstainers, Students and others. By Julius Erfurt, Manager of a Paper Mill. Translated into English and Edited with Additions by Julius Hübner, F.C.S., Lecturer on Papermaking at the Manchester Municipal Technical School. With illustrations and **157 patterns of paper dyed in the pulp.** Royal 8vo, 180 pp. Price 15s. net. (Post free, 15s. 6d. home; 16s. 6d. abroad).

THE PAPER MILL CHEMIST. By Henry P. Stevens, M.A., Ph.D., F.I.C. Royal 12mo. 60 illustrations. 300 pp. Price 7s. 6d. net. (Post free, 7s. 9d. home; 7s. 10d. abroad.)

THE TREATMENT OF PAPER FOR SPECIAL PURPOSES. By L.E. Andés. Translated from the German. Crown 8vo. 48 Illustrations. 250 pp. Price 6s. net. (Post free, 6s. 4d. home; 6s. 6d. abroad.)

(Enamelling on Metal.)

ENAMELS AND ENAMELLING. For Enamel Makers, Workers in Gold and Silver, and Manufacturers of Objects of Art. By Paul Randau. Translated from the German. With Sixteen Illustrations. Demy 8vo. 180 pp. Price 10s. 6d. net. (Post free, 10s. 10d. home; 11s. abroad.)

THE ART OF ENAMELLING ON METAL. By W. Norman Brown. Twenty-eight Illustrations. Crown 8vo. 60 pp. Price 2s. 6d. net. (Post free, 2s. 9d. home and abroad.)

[Pg 14]

(Textile and Dyeing Subjects.)

THE FINISHING OF TEXTILE FABRICS (Woollen, Worsted, Union and other Cloths). By Roberts Beaumont, M.Sc., M.I. Mech.E., Professor of Textile Industries, the University of Leeds; Author of "Colour in Woven Design"; "Woollen and Worsted Cloth Manufacture"; "Woven Fabrics at the World's Fair"; Vice-President of the Jury of Award at the Paris Exhibition, 1900; Inspector of Textile Institutes; Society of Arts Silver Medallist; Honorary Medallist of the City and Guilds of London Institute. With 150 Illustrations of Fibres, Yarns and Fabrics, also Sectional and other Drawings of Finishing Machinery Demy 8vo. 260 pp. Price 10s. 6d. net. (Post free, 10s. 10d. home; 11s. 3d. abroad.)

FIBRES USED IN TEXTILE AND ALLIED INDUSTRIES. By C. Ainsworth Mitchell, B.A. (Oxon.), F.I.C., and R.M. Prideaux, F.I.C. With 66 Illustrations specially drawn direct from the Fibres. Demy 8vo. 200 pp. Price 7s. 6d. net. (Post free, 7s. 10d. home; 8s. abroad.)

DRESSINGS AND FINISHINGS FOR TEXTILE FABRICS AND THEIR APPLICATION. Description of all the Materials used in Dressing Textiles: Their Special Properties, the preparation of Dressings and their employment in Finishing Linen, Cotton, Woollen and Silk Fabrics. Fireproof and Waterproof Dressings, together with the principal machinery employed. Translated from the Third German Edition of Friedrich Polleyn. Demy 8vo. 280 pp. Sixty Illustrations. Price 7s. 6d. net. (Post free, 7s. 10d. home; 8s. abroad.)

THE CHEMICAL TECHNOLOGY OF TEXTILE FIBRES; Their Origin, Structure, Preparation, Washing, Bleaching, Dyeing, Printing and Dressing. By Dr. Georg von Georgievics. Translated from the German by Charles Salter. 320 pp. Forty-seven Illustrations. Royal 8vo. Price 10s. 6d. net. (Post free, 11s. home; 11s. 3d. abroad.)

POWER-LOOM WEAVING AND YARN NUMBERING, According to Various Systems, with Conversion Tables. Translated from the German of Anthon Gruner. **With Twenty-six Diagrams in Colours.** 150 pp. Crown 8vo. Price 7s. 6d. net. (Post free, 7s. 9d. home; 8s. abroad.)

TEXTILE RAW MATERIALS AND THEIR CONVERSION INTO YARNS. (The Study of the Raw Materials and the Technology of the Spinning Process.) By Julius Zipser. Translated from German by Charles Salter. 302 Illustrations. 500 pp. Demy 8vo. Price 10s. 6d. net. (Post free, 11s. home; 11s. 6d. abroad.)

For contents of these books, see List II.

[Pg 15]

GRAMMAR OF TEXTILE DESIGN. By H. Nisbet, Weaving and Designing Master, Bolton Municipal Technical School. Demy 8vo. 280 pp. 490 Illustrations and Diagrams. Price 6s. net. (Post free, 6s. 4d. home; 6s. 6d. abroad.)

ART NEEDLEWORK AND DESIGN. POINT LACE. A Manual of Applied Art for Secondary Schools and Continuation Classes. By M.E. Wilkinson. Oblong quarto. With 22 Plates. Bound in Art Linen. Price 3s. 6d. net. (Post free, 3s. 10d. home; 4s. abroad.)

HOME LACE-MAKING. A Handbook for Teachers and Pupils. By M.E.W. Milroy. Crown 8vo. 64 pp. With 3 Plates and 9 Diagrams. Price 1s. net. (Post free, 1s. 3d. home; 1s. 4d. abroad.)

THE CHEMISTRY OF HAT MANUFACTURING. Lectures delivered before the Hat Manufacturers' Association. By Watson Smith, F.C.S., F.I.C. Revised and Edited by Albert Shonk. Crown 8vo. 132 pp. 16 Illustrations. Price 7s. 6d. net. (Post free, 7s. 9d. home; 7s. 10d. abroad.)

THE TECHNICAL TESTING OF YARNS AND TEXTILE FABRICS. With Reference to Official Specifications. Translated from the German of Dr. J. Herzfeld. Second Edition. Sixty-nine Illustrations. 200 pp. Demy 8vo. Price 10s. 6d. net. (Post free, 10s. 10d. home; 11s. abroad.)

DECORATIVE AND FANCY TEXTILE FABRICS. By R.T. Lord. For Manufacturers and Designers of Carpets, Damask, Dress and all Textile Fabrics. 200 pp. Demy 8vo. 132 Designs and Illustrations. Price 7s. 6d. net. (Post free, 7s. 10d. home; 8s. abroad.)

THEORY AND PRACTICE OF DAMASK WEAVING. By H. Kinzer and K. Walter. Royal 8vo. Eighteen Folding Plates. Six Illustrations. Translated from the German. 110 pp. Price 8s. 6d. net. (Post free, 9s. home; 9s. 6d. abroad.)

FAULTS IN THE MANUFACTURE OF WOOLLEN GOODS AND THEIR PREVENTION. By Nicolas Reiser. Translated from the Second German Edition. Crown 8vo. Sixty-three Illustrations. 170 pp. Price 5s. net. (Post free, 5s. 4d. home; 5s. 6d. abroad.)

SPINNING AND WEAVING CALCULATIONS, especially relating to Woollens. From the German of N. Reiser. Thirty-four Illustrations. Tables. 160 pp. Demy 8vo. 1904. Price 10s. 6d. net. (Post free, 10s. 10d. home; 11s. abroad.)

[Pg 16]

WATERPROOFING OF FABRICS. By Dr. S. Mierzinski. Crown 8vo. 104 pp. 29 Illus. Price 5s. net. (Post free, 5s. 3d. home; 5s. 4d. abroad.)

HOW TO MAKE A WOOLLEN MILL PAY. By John Mackie. Crown 8vo. 76 pp. Price 3s. 6d. net. (Post free, 3s. 9d. home; 3s. 10d. abroad.)

YARN AND WARP SIZING IN ALL ITS BRANCHES. Translated from the German of Carl Kretschmar. Royal 8vo. 123 Illustrations. 150 pp. Price 10s. 6d. net. (Post free, 10s. 10d. home; 11s. abroad.)

(For "Textile Soaps and Oils" see p. 7.)

(Dyeing, Colour Printing, Matching and Dye-stuffs.)

THE COLOUR PRINTING OF CARPET YARNS. Manual for Colour Chemists and Textile Printers. By David Paterson, F.C.S. Seventeen Illustrations. 136 pp. Demy 8vo. Price 7s. 6d. net. (Post free, 7s. 10d. home; 8s. abroad.)

THE SCIENCE OF COLOUR MIXING. A Manual intended for the use of Dyers, Calico Printers and Colour Chemists. By David Paterson, F.C.S. Forty-one Illustrations. **Five Coloured Plates, and Four Plates showing Eleven Dyed Specimens Of Fabrics.** 132 pp. Demy 8vo. Price 7s. 6d. net. (Post free, 7s. 10d. home; 8s. abroad.)

DYERS' MATERIALS: An Introduction to the Examination, Evaluation and Application of the most important Substances used in Dyeing, Printing, Bleaching and Finishing. By Paul Heerman, Ph.D. Translated from the German by A.C. Wright, M.A. (Oxon).,

B.Sc. (Lond.). Twenty-four Illustrations. Crown 8vo. 150 pp. Price 5s. net. (Post free, 5s. 4d. home; 5s. 6d. abroad.)

COLOUR MATCHING ON TEXTILES. A Manual intended for the use of Students of Colour Chemistry, Dyeing and Textile Printing. By David Paterson, F.C.S. Coloured Frontispiece. Twenty-nine Illustrations and **Fourteen Specimens of Dyed Fabrics.** Demy 8vo. 132 pp. Price 7s. 6d. net. (Post free, 7s. 10d. home; 8s. abroad.)

COLOUR: A HANDBOOK OF THE THEORY OF COLOUR. By George H. Hurst, F.C.S. **With Ten Coloured Plates** and Seventy-two Illustrations. 160 pp. Demy 8vo. Price 7s. 6d. net. (Post free, 7s. 10d. home; 8s. abroad.)

For contents of these books, see List II .

[Pg 17]

Reissue of

THE ART OF DYEING WOOL, SILK AND COTTON. Translated from the French of M. Hellot, M. Macquer and M. Le Pileur D'Apligny. First Published in English in 1789. Six Plates. Demy 8vo. 446 pp. Price 5s. net. (Post free, 5s. 6d. home; 6s. abroad.)

THE CHEMISTRY OF DYE-STUFFS. By Dr. Georg Von Georgievics. Translated from the Second German Edition. 412 pp. Demy 8vo. Price 10s. 6d. net. (Post free, 11s. home; 11s. 6d. abroad.)

THE DYEING OF COTTON FABRICS: A Practical Handbook for the Dyer and Student. By Franklin Beech, Practical Colourist and Chemist. 272 pp. Forty-four Illustrations of Bleaching and Dyeing Machinery. Demy 8vo. Price 7s. 6d. net. (Post free, 7s. 10d. home; 8s. abroad.)

THE DYEING OF WOOLLEN FABRICS. By Franklin Beech, Practical Colourist and Chemist. Thirty-three Illustrations. Demy 8vo. 228 pp. Price 7s. 6d. net. (Post free, 7s. 10d. home; 8s. abroad.)

(Silk Manufacture.)

SILK THROWING AND WASTE SILK SPINNING. By Hollins Rayner. Demy 8vo. 170 pp. 117 Illus. Price 5s. net. (Post free, 5s. 4d. home; 5s. 6d. abroad.)

(Bleaching and Bleaching Agents.)

A PRACTICAL TREATISE ON THE BLEACHING OF LINEN AND COTTON YARN AND FABRICS. By L. Tailfer, Chemical and Mechanical Engineer. Translated from the French by John Geddes McIntosh. Demy 8vo. 303 pp. Twenty Illus. Price 12s. 6d. net. (Post free, 13s. home; 13s. 6d. abroad.)

MODERN BLEACHING AGENTS AND DETERGENTS. By Professor Max Bottler. Translated from the German. Crown 8vo. 16 Illustrations. 160 pages. Price 5s. net. (Post free, 5s. 3d. home; 5s. 6d. abroad.)

(Cotton Spinning and Combing.)

COTTON SPINNING (First Year). By Thomas Thornley, Spinning Master, Bolton Technical School. 160 pp. Eighty-four Illustrations. Crown 8vo. Second Impression. Price 3s. net. (Post free, 3s. 4d. home; 3s. 6d. abroad.)

COTTON SPINNING (Intermediate, or Second Year). By Thomas Thornley. Second Impression. 180 pp. Seventy Illustrations. Crown 8vo. Price 5s. net. (Post free, 5s. 4d. home: 5s. 6d. abroad.)

[Pg 18]

COTTON SPINNING (Honours, or Third Year). By Thomas Thornley. 216 pp Seventy-four Illustrations. Crown 8vo. Second Edition. Price 5s. net. (Post free, 5s. 4d. home; 5s. 6d. abroad.)

COTTON COMBING MACHINES. By Thos. Thornley, Spinning Master, Technical School, Bolton. Demy 8vo. 117 Illustrations. 300 pp. Price 7s. 6d. net. (Post free, 8s. home; 8s. 6d. abroad.)

COTTON WASTE: Its Production, Characteristics, Regulation, Opening, Carding, Spinning and Weaving. By Thomas Thornley. Demy 8vo. About 300 pages. [*In the press.*

THE RING SPINNING FRAME: GUIDE FOR OVERLOOKERS AND STUDENTS. By N. Booth. Crown 8vo. 76 pages. Price 3s. net. (Post free, 3s. 3d. home; 3s. 6d. abroad.) [*Just published.*

(Flax, Hemp and Jute Spinning.)

MODERN FLAX, HEMP AND JUTE SPINNING AND TWISTING. A Practical Handbook for the use of Flax, Hemp and Jute Spinners, Thread, Twine and Rope Makers. By Herbert R. Carter, Mill Manager, Textile Expert and Engineer, Examiner in Flax Spinning to the City and Guilds of London Institute. Demy 8vo. 1907. With 92 Illustrations. 200 pp. Price 7s. 6d. net. (Post free, 7s. 9d. home; 8s abroad.)

(Collieries and Mines.)

RECOVERY WORK AFTER PIT FIRES. By Robert Lamprecht, Mining Engineer and Manager. Translated from the German. Illustrated by Six large Plates, containing Seventy-six Illustrations. 175 pp. Demy 8vo. Price 10s. 6d. net. (Post free, 10s. 10d. home; 11s. abroad.)

VENTILATION IN MINES. By Robert Wabner, Mining Engineer. Translated from the German. Royal 8vo. Thirty Plates and Twenty-two Illustrations. 240 pp. Price 10s. 6d. net. (Post free, 11s. home; 11s. 3d. abroad.)

HAULAGE AND WINDING APPLIANCES USED IN MINES. By Carl Volk. Translated from the German. Royal 8vo. With Six Plates and 148 Illustrations. 150 pp. Price 8s. 6d. net. (Post free, 9s. home; 9s. 3d. abroad.)

For contents of these books, see List III.

[Pg 19]

THE ELECTRICAL EQUIPMENT OF COLLIERIES. By W. Galloway Duncan, Electrical and Mechanical Engineer, Member of the Institution of Mining Engineers, Head of the Government School of Engineering, Dacca, India; and David Penman, Certificated Colliery Manager, Lecturer in Mining to Fife County Committee. Demy 8vo. 310 pp. 155 Illustrations and Diagrams. Price 10s. 6d. net. (Post free, 11s. home; 11s. 3d. abroad.)

(Dental Metallurgy.)

DENTAL METALLURGY: MANUAL FOR STUDENTS AND DENTISTS. By A.B. Griffiths, Ph.D. Demy 8vo. Thirty-six Illustrations. 200 pp. Price 7s. 6d. net. (Post free, 7s. 10d. home; 8s. abroad.)

(Engineering, Smoke Prevention and Metallurgy.)

THE PREVENTION OF SMOKE. Combined with the Economical Combustion of Fuel. By W.C. Popplewell, M.Sc., A.M. Inst., C.E., Consulting Engineer. Forty-six Illustrations. 190 pp. Demy 8vo. Price 7s. 6d. net. (Post free, 7s. 10d. home; 8s. 3d. abroad.)

GAS AND COAL DUST FIRING. A Critical Review of the Various Appliances Patented in Germany for this purpose since 1885. By Albert Pütsch. 130 pp. Demy 8vo. Translated from the German. With 103 Illustrations. Price 5s. net. (Post free, 5s. 4d. home; 5s. 6d. abroad.)

THE HARDENING AND TEMPERING OF STEEL IN THEORY AND PRACTICE. By Fridolin Reiser. Translated from the German of the Third Edition. Crown 8vo. 120 pp. Price 5s. net. (Post free, 5s. 3d. home; 5s. 4d. abroad.)

SIDEROLOGY: THE SCIENCE OF IRON (The Constitution of Iron Alloys and Slags). Translated from German of Hanns Freiherr v. Jüptner. 350 pp. Demy 8vo. Eleven Plates and Ten Illustrations. Price 10s. 6d. net. (Post free, 11s. home; 11s. 6d. abroad.)

EVAPORATING, CONDENSING AND COOLING APPARATUS. Explanations, Formulæ and Tables for Use in Practice. By E. Hausbrand, Engineer. Translated by A.C. Wright, M.A. (Oxon.), B.Sc., (Lond.). With Twenty-one Illustrations and Seventy-six Tables. 400 pp. Demy 8vo. Price 10s. 6d. net. (Post free, 11s. home; 11s. 6d. abroad.)

[Pg 20]

(The "Broadway" Series of Engineering Handbooks.)

Volume I.—**REINFORCED CONCRETE.** By Ewart S. Andrews, B.Sc. Eng. (Lond.). [*In the press.*

Volume II.—**GAS AND OIL ENGINES.** [*In the press.*

Volume III. – **STRUCTURAL STEEL AND IRON WORK.** [*In the press.*

Volume IV. – **TOOTHED GEARING.** By G.T. White, B.Sc. (Lond.). [*In the press.*

Volume V. – **STEAM TURBINES:** Their Theory and Construction. [*In the press.*

(Sanitary Plumbing, Electric Wiring, Metal Work, etc.)

EXTERNAL PLUMBING WORK. A Treatise on Lead Work for Roofs. By John W. Hart, R.P.C. 180 Illustrations. 272 pp. Demy 8vo. Second Edition Revised. Price 7s. 6d. net. (Post free. 7s. 10d. home; 8s. abroad.)

HINTS TO PLUMBERS ON JOINT WIPING, PIPE BENDING AND LEAD BURNING. Third Edition, Revised and Corrected, By John W. Hart, R.P.C. 184 Illustrations. 313 pp. Demy 8vo. Price 7s. 6d. net. (Post free, 8s. home; 8s. 6d. abroad.)

SANITARY PLUMBING AND DRAINAGE. By John W. Hart. Demy 8vo. With 208 Illustrations. 250 pp. 1904. Price 7s. 6d. net. (Post free, 7s. 10d. home; 8s. abroad.)

ELECTRIC WIRING AND FITTING. By Sydney F. Walker, R.N., M.I.E.E., M.I.Min.E., A.M.Inst.C.E., etc., etc. Crown 8vo. 150 pp. With Illustrations and Tables. Price 5s. net. (Post free, 5s. 3d. home; 5s. 6d. abroad.)

THE PRINCIPLES AND PRACTICE OF DIPPING, BURNISHING, LACQUERING AND BRONZING BRASS WARE. By W. Norman Brown. 48 pp. Crown 8vo. Price 3s. net. (Post free, 3s. 3d. home and abroad.) [*Just published.*

THE DEVELOPMENT OF THE INCANDESCENT ELECTRIC LAMPS. By G. Basil Barham, A.M.I.E.E. Illustrated. Demy 8vo. 196 pp. [*In the press.*

For contents of these books, see List I.

[Pg 21]

WIRING CALCULATIONS FOR ELECTRIC LIGHT AND POWER INSTALLATIONS. A Practical Handbook containing

Wiring Tables, Rules, and Formulæ for the Use of Architects, Engineers, Mining Engineers, and Electricians, Wiring Contractors and Wiremen, etc. By G. Lummis Paterson. Crown 8vo. Twenty-two Illustrations. 100 pp. [*In the press.*

A HANDBOOK ON JAPANNING AND ENAMELLING FOR CYCLES, BEDSTEADS, TINWARE, ETC. By William Norman Brown. 52 pp. and Illustrations. Crown 8vo. Price 2s. net. (Post free, 2s. 3d. home and abroad.)

THE PRINCIPLES OF HOT WATER SUPPLY. By John W. Hart, R.P.C. With 129 Illustrations. 177 pp. Demy 8vo. Price 7s. 6d. net. (Post free, 7s. 10d. home; 8s. abroad.)

(Brewing and Botanical.)

HOPS IN THEIR BOTANICAL, AGRICULTURAL AND TECHNICAL ASPECT, AND AS AN ARTICLE OF COMMERCE. By Emmanuel Gross, Professor at the Higher Agricultural College, Tetschen-Liebwerd. Translated from the German. Seventy-eight Illustrations. 340 pp. Demy 8vo. Price 10s. 6d. net. (Post free, 11s. home; 11s 6d. abroad.)

A BOOK ON THE DISEASES OF PLANTS, FUNGICIDES AND INSECTICIDES, ETC. Demy 8vo. About 500 pp. [*In the press.*

(Wood Products, Timber and Wood Waste.)

WOOD PRODUCTS: DISTILLATES AND EXTRACTS. By P. Dumesny, Chemical Engineer, Expert before the Lyons Commercial Tribunal, Member of the International Association of Leather Chemists; and J. Noyer. Translated from the French by Donald Grant. Royal 8vo. 320 pp. 103 Illustrations and Numerous Tables. Price 10s. 6d. net. (Post free, 11s. home; 11s. 6d. abroad.)

TIMBER: A Comprehensive Study of Wood in all its Aspects (Commercial and Botanical), showing the different Applications and Uses of Timber in Various Trades, etc. Translated from the French of Paul Charpentier. Royal 8vo. 437 pp. 178 Illustrations. Price 12s. 6d. net. (Post free, 13s. home; 14s. abroad.)

[Pg 22]

THE UTILISATION OF WOOD WASTE. Translated from the German of Ernst Hubbard. Crown 8vo. 192 pp. Fifty Illustrations. Price 5s. net. (Post free, 5s. 4d. home; 5s. 6d. abroad.)

(See also Utilisation of Waste Products, p. 9.)

(Building and Architecture.)

ORNAMENTAL CEMENT WORK. By Oliver Wheatley. Demy 8vo. 83 Illustrations. 128 pp. Price 5s. net. (Post free, 5s. 4d. home; 5s. 6d. abroad.) [*Just published.*

THE PREVENTION OF DAMPNESS IN BUILDINGS; with Remarks on the Causes, Nature and Effects of Saline, Efflorescences and Dry-rot, for Architects, Builders, Overseers, Plasterers, Painters and House Owners. By Adolf Wilhelm Keim. Translated from the German of the second revised Edition by M.J. Salter, F.I.C., F.C.S. Eight Coloured Plates and Thirteen Illustrations. Crown 8vo. 115 pp. Price 5s. net. (Post free, 5s. 3d. home; 5s. 4d. abroad.)

HANDBOOK OF TECHNICAL TERMS USED IN ARCHITECTURE AND BUILDING, AND THEIR ALLIED TRADES AND SUBJECTS. By Augustine C. Passmore. Demy 8vo. 380 pp. Price 7s. 6d. net. (Post free, 8s. home; 8s. 6d. abroad.)

(Foods, Drugs and Sweetmeats.)

FOOD AND DRUGS. By E.J. Parry, B.Sc., F.I.C., F.C.S. Volume I. The Analysis of Food and Drugs (Chemical and Microscopical). Royal 8vo. 724 pp. Price 21s. net. (Post free, 21s. 8d. home; 22s. abroad.) Volume II. The Sale of Food and Drugs Acts, 1875-1907. Royal 8vo. 184 pp. Price 7s. 6d. net. (Post free, 7s. 10d. home; 8s. abroad.) [*Just published.*

THE MANUFACTURE OF PRESERVED FOODS AND SWEETMEATS. By A. Hausner. With Twenty-eight Illustrations. Translated from the German of the third enlarged Edition. Crown 8vo. 225 pp. Price 7s. 6d. net. (Post free, 7s. 9d. home; 7s. 10d. abroad.)

RECIPES FOR THE PRESERVING OF FRUIT, VEGETABLES AND MEAT. By E. Wagner. Translated from the German. Crown

8vo. 125 pp. With 14 Illustrations. Price 5s. net. (Post free, 5s. 3d. home; 5s. 4d. abroad.)

For contents of these books, see List III.

[Pg 23]

(Dyeing Fancy Goods.)

THE ART OF DYEING AND STAINING MARBLE, ARTIFICIAL STONE, BONE, HORN, IVORY AND WOOD, AND OF IMITATING ALL SORTS OF WOOD. A Practical Handbook for the Use of Joiners, Turners, Manufacturers of Fancy Goods, Stick and Umbrella Makers, Comb Makers, etc. Translated from the German of D.H. Soxhlet, Technical Chemist. Crown 8vo. 168 pp. Price 5s. net. (Post free, 5s. 3d. home; 5s. 4d. abroad.)

(Celluloid.)

CELLULOID: Its Raw Material, Manufacture, Properties and Uses. A Handbook for Manufacturers of Celluloid and Celluloid Articles, and all Industries using Celluloid; also for Dentists and Teeth Specialists. By Dr. Fr. Böckmann, Technical Chemist. Translated from the Third Revised German Edition. Crown 8vo. 120 pp. With 49 Illustrations. Price 5s. net. (Post free, 5s. 3d. home; 5s. 4d. abroad.)

(Lithography, Printing and Engraving.)

PRACTICAL LITHOGRAPHY. By Alfred Seymour. Demy 8vo. With Frontispiece and 33 Illus. 120 pp. Price 5s. net. (Post free, 5s. 4d. home; 5s. 6d. abroad.)

PRINTERS' AND STATIONERS' READY RECKONER AND COMPENDIUM. Compiled by Victor Graham. Crown 8vo. 112 pp. 1904. Price 3s. 6d. net. (Post free, 3s. 9d. home; 3s. 10d. abroad.)

ENGRAVING FOR ILLUSTRATION. HISTORICAL AND PRACTICAL NOTES. By J. Kirkbride. 72 pp. Two Plates and 6 Illustrations. Crown 8vo. Price 2s. 6d. net. (Post free, 2s. 9d. home; 2s. 10d. abroad.)

(For Printing Inks, see p. 4.)

(Bookbinding.)

PRACTICAL BOOKBINDING. By Paul Adam. Translated from the German. Crown 8vo. 180 pp. 127 Illustrations. Price 5s. net. (Post free, 5s. 4d. home; 5s. 6d. abroad.)

(Sugar Refining.)

THE TECHNOLOGY OF SUGAR: Practical Treatise on the Modern Methods of Manufacture of Sugar from the Sugar Cane and Sugar Beet. By John Geddes McIntosh. Second Revised and Enlarged Edition. Demy 8vo. Fully Illustrated. 436 pp. Seventy-six Tables. 1906. Price 10s. 6d. net. (Post free, 11s. home; 11s. 6d. abroad.)

(See "Evaporating, Condensing, etc., Apparatus," p. 9.)

[Pg 24]

(Emery.)

EMERY AND THE EMERY INDUSTRY. Translated from the German of A. Haenig. Crown 8vo. 45 Illustrations. 110 pp. Price 5s. net. (Post free, 5s. 3d. home; 5s. 6d. abroad.) *[Just published.*

(Libraries and Bibliography.)

CLASSIFIED GUIDE TO TECHNICAL AND COMMERCIAL BOOKS. Compiled by Edgar Greenwood. Demy 8vo. 224 pp. 1904. Being a Subject-list of the Principal British and American Books in Print; giving Title, Author, Size, Date, Publisher and Price. Price 5s. net. (Post free, 5s. 4d. home; 5s. 6d. abroad.)

HANDBOOK TO THE TECHNICAL AND ART SCHOOLS AND COLLEGES OF THE UNITED KINGDOM. Containing particulars of nearly 1,000 Technical, Commercial and Art Schools throughout the United Kingdom. With full particulars of the courses of instruction, names of principals, secretaries, etc. Demy 8vo. 150 pp. Price 3s. 6d. net. (Post free, 3s. 10d. home; 4s. abroad.)

THE LIBRARIES, MUSEUMS AND ART GALLERIES YEAR BOOK, 1910-11. Being the Third Edition of Greenwood's "British

Library Year Book". Edited by Alex. J. Philip. Demy 8vo. 286 pp. Price 5s. net. (Post free, 5s. 4d. home; 5s. 6d. abroad.)

THE PLUMBING, HEATING AND LIGHTING ANNUAL FOR 1911. The Trade Reference Book for Plumbers, Sanitary, Heating and Lighting Engineers, Builders' Merchants, Contractors and Architects. Quarto. Bound in cloth and gilt lettered. Price 3s. net. (Post free, 3s. 4d. home; 3s. 8d. abroad.)

Including the translation of Hermann Kechnagel's "Kalender fur Gesundheits-Techniker," Handbook for Heating, Ventilating, and Domestic Engineers, of which Scott, Greenwood & Son have purchased the sole right for the English Language.

SCOTT, GREENWOOD & SON,
Technical Book and Trade Journal Publishers,
8 Broadway, Ludgate Hill,
London, E.C.

Telegraphic Address, "Printeries, London". Tel. No.: Bank 5403.
January, 1912.

www.ingramcontent.com/pod-product-compliance
Lightning Source LLC
Chambersburg PA
CBHW031421210526
45464CB00005B/1994